SpringerBriefs in Mathematics

SpringerBriefs in Mathematics showcases expositions in all areas of mathematics and applied mathematics. Manuscripts presenting new results or a single new result in a classical field, new field, or an emerging topic, applications, or bridges between new results and already published works, are encouraged. The series is intended for mathematicians and applied mathematicians.

BCAM SpringerBriefs

Editorial Board

Enrique Zuazua
BCAM—Basque Center for Applied Mathematics
& Ikerbasque
Bilbao, Basque Country, Spain

Irene Fonseca
Center for Nonlinear Analysis
Department of Mathematical Sciences
Carnegie Mellon University
Pittsburgh, USA

Juan J. Manfredi
Department of Mathematics
University of Pittsburgh
Pittsburgh, USA

Emmanuel Trélat
Laboratoire Jacques-Louis Lions
Institut Universitaire de France
Université Pierre et Marie Curie
CNRS, UMR, Paris

Xu Zhang
School of Mathematics
Sichuan University
Chengdu, China

BCAM *SpringerBriefs* aims to publish contributions in the following disciplines: Applied Mathematics, Finance, Statistics and Computer Science. BCAM has appointed an Editorial Board, who evaluate and review proposals.

Typical topics include: a timely report of state-of-the-art analytical techniques, bridge between new research results published in journal articles and a contextual literature review, a snapshot of a hot or emerging topic, a presentation of core concepts that students must understand in order to make independent contributions.

Please submit your proposal to the Editorial Board or to Francesca Bonadei, Executive Editor Mathematics, Statistics, and Engineering: francesca.bonadei@springer.com

More information about this series at http://www.springer.com/series/10030

Frédéric Jean

Control of Nonholonomic Systems: From Sub-Riemannian Geometry to Motion Planning

Frédéric Jean
UMA
ENSTA ParisTech
Palaiseau
France

ISSN 2191-8198   ISSN 2191-8201   (electronic)
ISBN 978-3-319-08689-7 ISBN 978-3-319-08690-3   (eBook)
DOI 10.1007/978-3-319-08690-3

Library of Congress Control Number: 2014944540

Springer Cham Heidelberg New York Dordrecht London

Printed on acid-free paper

Springer is part of Springer Science+Business Media (www.springer.com)

Preface

Nonholonomic systems are control systems which depend linearly on the control. Their underlying geometry is the sub-Riemannian geometry, which plays for these systems the same role as Euclidean geometry does for linear systems. In particular the usual notions of approximations at the first order, that are essential for control purposes, have to be defined in terms of this geometry. The aim of these notes is to present these notions of approximation and their application to the motion planning problem for nonholonomic systems.

The notes are divided into three chapters and two appendices. In Chap. 1, we introduce the basic definitions on nonholonomic systems and sub-Riemannian geometry, and give the main result on controllability, namely the Chow–Rashevsky Theorem. Chapter 2 provides a detailed exposition of the notions of first-order approximation, including nonholonomic orders, privileged coordinates, nilpotent approximations, and distance estimates such as the Ball-Box Theorem. As an application, we show how these notions allow us to describe the tangent structure to a Carnot-Carathéodory space (the metric space defined by a sub-Riemannian distance). The chapter ends with the presentation of desingularization procedures that are necessary to recover uniformity in approximations and distance estimates. Chapter 3 is devoted to the motion planning problem for nonholonomic systems. We show in particular how to apply the tools from sub-Riemannian geometry in order to give solutions to this problem, first in the case where the system is nilpotent, and then in the general case. An overview of the existing methods for nonholonomic motion planning conclude this chapter. Finally, we present some results on composition of flows in connection with the Campbell-Hausdorff formula in Appendix A, and some complements on the different systems of privileged coordinates in Appendix B.

From the point of view of the sub-Riemannian geometry, this book is intended to be complementary to that of Ludovic Rifford in the same collection [1]. As a consequence, the subjects that are extensively talked about in the latter (for instance sub-Riemannian geodesics) are not discussed here.

Notice finally that the main theoretical part about controllability and first-order theory is self-contained, all the results being proved. However for some applications, we took the liberty of stating some results without demonstration. They concern either developments beyond the scope of these notes (Theorem 2.4 called the Uniform Ball-Box Theorem, Theorem 2.5 on the metric tangent cone) or technical results on algorithmic procedures (Theorem 2.9 on the desingularization procedure, the fact that the formula (3.17) on sinusoidal controls may be inverted).

These notes grew out of a series of lectures given at the *Trimester on Dynamical and Control Systems* in Trieste in 2003, and more recently at the CIMPA Schools *Géométrie sous-riemannienne* in Beirut, Lebanon, in 2012, and *Contrôle géométrique, stochastique et des équations aux dérivées partielles* in Tlemcen, Algeria, in 2014. I am most grateful to the organizers of these events, Andrei Agrachev and Ugo Boscain for the first one, Fernand Pelletier, Ali Fardoun, and Mohamad Mehdi for the second one, and Sidi Mohammed Bouguima, Benmiloud Mebkhout, and Yacine Chitour for the third one. The materials of the third chapter mostly come from a collaboration with Yacine Chitour and Ruixing Long, during the Ph.D. thesis of the latter. This book is also thanks to them.

Paris, April 2014   Frédéric Jean

Reference

1. Rifford, L.: *Sub-Riemannian geometry and optimal transport*. SpringerBriefs in Mathematics. Springer International Publishing, New york (2014)

Acknowledgments

This work was supported by iCODE (Institute for Control and Decision), research project of the IDEX Paris- Saclay, by the ANR project *GCM*, program "Blanche," project number NT09_504490, and by the Commission of the European Communities under the 7th Framework Programme Marie Curie Initial Training Network (FP7-PEOPLE-2010-ITN), project SADCO, contract number 264735.

Contents

Chapter 1
Geometry of Nonholonomic Systems

Abstract This first chapter presents the basic definitions and results on nonholonomic systems and sub-Riemannian distance. We begin in Sect. 1.1 with a discussion on linearization of control systems, which is an underlying theme of this book. We then introduce in Sect. 1.2 the definition of nonholonomic systems and of the associated sub-Riemannian distances, that we distinguish from the ones associated with sub-Riemannian manifolds defined in Sect. 1.3. Finally we state and prove in Sect. 1.4 Chow-Rashevsky's theorem as well as rough estimates for sub-Riemannian distances.

Keywords Control theory · Sub-Riemannian geometry · Nonholonomic systems

Throughout these notes we work in a smooth n-dimensional manifold M.

1.1 Introduction

Let us introduce these notes by some considerations on control theory. Consider a nonlinear control system in $\mathbb{R}^n$,

$$\dot{x} = f(x, u),$$

where $x \in \mathbb{R}^n$ is the state and $u \in \mathbb{R}^m$ is the control. Given a control law $u(t)$, $t \in [0, T]$, a trajectory associated with $u(\cdot)$ is defined as a solution of the nonautonomous ordinary differential equation $\dot{x} = f(x, u(t))$. The first question is the one of the controllability: for any pair of points, does there exist a control law $u(t)$, $t \in [0, T]$, such that the associated trajectory joins one point to the other? In the case where the answer is positive, next issues are notably the motion planning (i.e. find a solution $u(\cdot)$ to the previous question) and the stabilization (i.e. design the control as a function $u(t) = k(x(t))$ of the state in such a way that the resulting differential equation $\dot{x} = f(x, k(x))$ is stable).

The usual way to deal with these problems locally is to use a first-order approximation of the system. The underlying idea is the following. Consider a pair $(\bar{x}, \bar{u}) \in \mathbb{R}^{m+n}$ such that $f(\bar{x}, \bar{u}) = 0$. The linearized system around this equilibrium

© The Author(s) 2014

F. Jean, *Control of Nonholonomic Systems: From Sub-Riemannian Geometry to Motion Planning*, SpringerBriefs in Mathematics, DOI 10.1007/978-3-319-08690-3_1

pair is defined to be the linear control system,

$$\dot{\delta x} = \frac{\partial f}{\partial x}(\bar{x}, \bar{u})\delta x + \frac{\partial f}{\partial u}(\bar{x}, \bar{u})\delta u,$$

where $\delta x \in \mathbb{R}^n$ is the state and $\delta u \in \mathbb{R}^m$ is the control. If this linearized system is controllable, so is the nonlinear one near $\bar{x}$. In this case the solutions to the motion planning and stabilization problems for the linearized system may be used to construct solutions of the corresponding problems for the nonlinear system. Thus, locally, the study of the control system amounts to the one of the linearized system.

Assume now that the control system depends linearly on u [but a priori not on (x, u)], that is,

$$\dot{x} = \sum_{i=1}^{m} u_i X_i(x). \tag{1.1}$$

Does the strategy above apply? For every $\bar{x} \in \mathbb{R}^n$, the pair $(\bar{x}, 0)$ is an equilibrium pair and the corresponding linearized system is

$$\dot{\delta x} = \sum_{i=1}^{m} \delta u_i X_i(\bar{x}), \quad \delta x \in \mathbb{R}^n, \ \delta u \in \mathbb{R}^m.$$

For this linearized system, the reachable set from a point δx is obviously the affine subset $\delta x + \Delta(\bar{x})$, where $\Delta(\bar{x}) = \mathrm{span}\, \{X_1(\bar{x}), \ldots, X_m(\bar{x})\}$. We distinguish two cases depending on the dimension of $\Delta(\bar{x})$:

- if $\dim \Delta(\bar{x}) = n$, then the linearized system is controllable and the strategy by linearization may be usefully applied. Note that in this case the original system (1.1) is also locally controllable since it admits as trajectory every C^1 curve near $\bar{x}$;
- if $\dim \Delta(\bar{x}) < n$, then the linearized system is not controllable. However System (1.1) may be controllable (and generically it is), as we will see in Sect. 1.4. Hence the strategy by linearization does not apply, the linearized system does not reflect the local behaviour of the nonlinear system.

Nonholonomic systems are precisely the systems of the form (1.1) which belong to the second category. The fact that for such systems the linearized system is useless, may be understood as follows. The linearization is a first-order approximation with respect to a Euclidean (or a Riemannian) distance. However for nonholonomic systems the underlying distance is a sub-Riemannian one and it behaves very differently from a Euclidean one. Thus, the local behaviour should be understood through the study of a first-order approximation with respect to this sub-Riemannian distance, not through the linearized system.

Let us illustrate these ideas with an example, the Brockett integrator. Consider the nonholonomic system in $\mathbb{R}^3$,

$$\dot{x} = u_1, \ \dot{y} = u_2, \ \dot{z} = xu_2, \tag{1.2}$$

defined by the vector fields $X_1(q) = (1, 0, 0)$ and $X_2(q) = (0, 1, x)$, where $q = (x, y, z)$. The linearized system around $(q = 0, u = 0)$ writes as,

$$\dot{\delta x} = \delta u_1, \ \dot{\delta y} = \delta u_2, \ \dot{\delta z} = 0,$$

and is non controllable since no motion in the δz direction is allowed. On the other hand, (1.2) is clearly controllable; in particular, one can move in the z direction by using the following control law,

$$u^*(t) = \begin{cases} (1, 0) & \text{for } t \in [0, \varepsilon], \\ (0, 1) & \text{for } t \in [\varepsilon, 2\varepsilon], \\ (-1, 0) & \text{for } t \in [2\varepsilon, 3\varepsilon], \\ (0, -1) & \text{for } t \in [3\varepsilon, 4\varepsilon], \end{cases}$$

which allows to steer the system from 0 to $q^1 = (0, 0, \varepsilon^2)$ in time 4ε. The fact that $\|q^1\|$ is of order 2 in function of the time, and not of order 1, explains why the control $u^*(\cdot)$ has no effect on the linearized system, and so why the latter is not controllable.

Let us try to quantify the relation between the duration of a motion and the attained point. For every $q \in \mathbb{R}^3$ we denote by $T(q)$ the minimal time needed to go from 0 to q with controls such that $\|u(t)\| \leq 1$. By using the control $u^*(\cdot)$ on the one hand, and direct upper bounds on Eq. (1.2) on the other hand, we easily obtain

$$\frac{1}{3} \left(|x| + |y| + |z|^{1/2} \right) \leq T(q) \leq 4 \left(|x| + |y| + |z|^{1/2} \right).$$

Thus $T(q)$ has to be compared with the *weighted* pseudo-norm $|x| + |y| + |z|^{1/2}$ and not with the usual Euclidean norm, and first-order approximations should be taken with respect to such a pseudo-norm.

We introduce now in a proper way nonholonomic systems and sub-Riemannian distances. We will come back to the notions of first-order approximations and weighted pseudo-norms in Chap. 2.

1.2 Nonholonomic Systems and Sub-Riemannian Distances

Definition 1.1 A *nonholonomic system* on M is a control system which is of the form

$$\dot{q} = u_1 X_1(q) + \cdots + u_m X_m(q), \quad q \in M, \quad u = (u_1, \ldots, u_m) \in \mathbb{R}^m, \tag{1.3}$$

where $m > 1$ is an integer and $X_1, \ldots, X_m$ are C^∞ vector fields on M.

The system (1.3) determines a family of vector spaces,

$$\Delta(q) = \text{span}\,\{X_1(q), \ldots, X_m(q)\} \subset T_q M, \quad q \in M.$$

The dimension of $\Delta(q)$ is a function of q, and may be non constant. If it is constant, Δ defines a distribution on M, that is, a subbundle of TM. Although we deal in general with systems such that the dimension of $\Delta(q)$ is smaller than n, we do not exclude the particular case where $\Delta(q) = T_q M$ for every $q \in M$. Strictly speaking in the latter case the system should be called a holonomic system, but for convenience we keep the vocable nonholonomic whatever the dimension of $\Delta(q)$.

To give a meaning to the control system above, we have to define what are its solutions, or trajectories.

Definition 1.2 A *trajectory of* (1.3) is a path $\gamma : [0, T] \to M$ for which there exists a function $u(\cdot) \in L^1([0, T], \mathbb{R}^m)$ such that γ is a solution of the ordinary differential equation,

$$\dot{q}(t) = \sum_{i=1}^{m} u_i(t) X_i(q(t)), \quad \text{for a.e. } t \in [0, T].$$

Such a function $u(\cdot)$ is called a *control* associated with γ.

In particular, every trajectory is an absolutely continuous path γ on M such that $\dot{\gamma}(t) \in \Delta(\gamma(t))$ for almost every $t \in [0, T]$.

Example 1.1 (*unicycle*) The most typical example of nonholonomic system is the simplified kinematic model of a unicycle. In this model, a configuration $q = (x, y, \theta)$ of the unicycle is described by the planar coordinates (x, y) of the contact point of the wheel with the ground, and by the angle θ of orientation of the wheel with respect to the x-axis. The space of configurations is then the manifold $\mathbb{R}^2 \times \mathscr{S}^1$.

The wheel is subject to the constraint of rolling without slipping, which writes as $\dot{x} \sin\theta - \dot{y}\cos\theta = 0$, or, equivalently as $\dot{q} \in \ker \omega(q)$, where ω is the one-form $\sin\theta\, dx - \cos\theta\, dy$. Hence the set Δ defined by the system is $\ker \omega$.

Choosing as controls the tangential velocity u_1 and the angular velocity u_2, we obtain the nonholonomic system $\dot{q} = u_1 X_1(q) + u_2 X_2(q)$ on $\mathbb{R}^2 \times \mathscr{S}^1$, where $X_1 = \cos\theta\, \partial_x + \sin\theta\, \partial_y$, and $X_2 = \partial_\theta$.

Let us mention here a few properties of the trajectories of (1.3) (for more details, see [4]).

- Fix $p \in M$ and $T > 0$. For every control $u(\cdot) \in L^1([0, T], \mathbb{R}^m)$, there exists $\tau \in (0, T]$ such that the Cauchy problem

$$\begin{cases} \dot{q}(t) = \sum_{i=1}^{m} u_i(t) X_i(q(t)) & \text{for a.e. } t \in [0, \tau], \\ q(0) = p, \end{cases} \tag{1.4}$$

has a unique solution denoted by γ_u or $\gamma(\cdot\,; p, u)$. It is called the *trajectory issued from p associated with u*.

- Any time-reparameterization of a trajectory is still a trajectory: if $\gamma: [0, T] \to M$ is a trajectory associated with a control u, and $\alpha: [0, S] \to [0, T]$ is a C^1-diffeomorphism, then $\gamma \circ \alpha: [0, S] \to M$ is a trajectory associated with the control $\alpha'(s)u(\alpha(s))$. In particular, one can reverse time along γ: the resulting path $\gamma(T - s)$, $s \in [0, T]$, is a trajectory associated with the control $-u(T - s)$.

Sub-Riemannian Distance

A nonholonomic system induces a distance on M in the following way. We first define the *sub-Riemannian metric* associated with (1.3) to be the function $g: TM \to \overline{\mathbb{R}}$ given by

$$g(q, v) = \inf \left\{ u_1^2 + \cdots + u_m^2 : \sum_{i=1}^{m} u_i X_i(q) = v \right\}, \tag{1.5}$$

for $q \in M$ and $v \in T_q M$, where we adopt the convention that $\inf \emptyset = +\infty$. This function g satisfies:

- if $v \notin \Delta(q)$, then $g(q, v) = +\infty$;
- if $v \in \Delta(q)$, then the infimum in (1.5) is attained at a unique value $u^* \in \mathbb{R}^m$, and thus $g(q, v) = \|u^*\|^2$ where $\| \cdot \|$ denotes the Euclidean norm on $\mathbb{R}^m$.

Such a metric allows to define a distance in the same way as in Riemannian geometry.

Definition 1.3 The *length* of an absolutely continuous path $\gamma(t)$, $t \in [0, T]$, is

$$\text{length}(\gamma) = \int_0^T \sqrt{g(\gamma(t), \dot{\gamma}(t))}\, dt$$

(the integral is well-defined since $g(\gamma(t), \dot{\gamma}(t))$ is measurable, being the composition of the lower semi-continuous function g with measurable functions). The *sub-Riemannian distance* on M associated with the nonholonomic system (1.3) is defined by

$$d(p, q) = \inf \text{length}(\gamma),$$

where the infimum is taken over all absolutely continuous paths γ joining p to q.

Consider an absolutely continuous path γ having finite length. It necessarily satisfies $\dot{\gamma}(t) \in \Delta(\gamma(t))$ for almost every $t \in [0, T]$. Denote by $u^*(t)$, $t \in [0, T]$, the measurable function defined by $g(\gamma(t), \dot{\gamma}(t)) = \|u^*(t)\|^2$. The finiteness of $\text{length}(\gamma)$ implies that $u^*(\cdot)$ belongs to $L^1([0, T], \mathbb{R}^m)$, and therefore that γ is a trajectory of (1.3) having $u^*(\cdot)$ as an associated control.

Thus only trajectories of (1.3) may have a finite length. As a consequence, if no trajectory connects p with q, then $d(p, q) = +\infty$. We will see below in Corollary 1.1 that, under an extra assumption on the nonholonomic system, d is actually finite and satisfies the properties of a distance function.

Remark 1.1 If the rank of $X_1, \ldots, X_m$ is constant and equal to m on M, every trajectory γ is associated with a unique control, otherwise different controls are associated with γ. However it results from the discussion above that the length of γ is equal to the L^1 norm $\|u^*\|_{L^1}$ of the unique control $u^*(\cdot)$ defined by $g\left(\gamma(t), \dot{\gamma}(t)\right) = \|u^*(t)\|^2$. We sometimes refer to $u^*(\cdot)$ as *the* control associated with γ. Note that length$(\gamma) = \|u^*\|_{L^1}$ is also the minimum of $\|u\|_{L^1}$ over all controls $u(\cdot)$ associated with γ.

An important feature of the length of a path is that it is independent of the parametrization of the path. As a consequence, the sub-Riemannian distance $d(p, q)$ may also be understood as the minimal time needed for the nonholonomic system to go from p to q with bounded controls, that is,

$$d(p, q) = \inf \left\{ T \geq 0 : \begin{array}{c} \exists \text{ a trajectory } \gamma_u : [0, T] \to M \text{ s.t.} \\ \gamma_u(0) = p, \ \gamma_u(T) = q, \\ \text{and } \|u(t)\| \leq 1 \text{ for a.e. } t \in [0, T] \end{array} \right\}. \tag{1.6}$$

This formulation justifies the assertion made in Sect. 1.1: for nonholonomic systems, first-order approximations with respect to the time should be understood as first-order approximations with respect to the sub-Riemannian distance.

Another consequence of (1.6) is that $d(p, q)$ is the solution of a time-optimal control problem. It then results from standard existence theorems (see for instance [2] or [4]) that, when p and q are sufficiently close and $d(p, q) < \infty$, there exists a trajectory γ connecting p with q such that

$$\text{length}(\gamma) = d(p, q).$$

Such a trajectory is called a *minimizing trajectory*.

Remark 1.2 Any reparameterization of a minimizing trajectory is also minimizing. Therefore any pair of close enough points can be joined by a minimizing trajectory *of velocity one*, that is, a trajectory γ such that $g(\gamma(t), \dot{\gamma}(t)) = 1$ for a.e. t. As a consequence, there exists a control $u(\cdot)$ associated with γ such that $\|u(t)\| = 1$ a.e. Every sub-arc of such a trajectory γ is also clearly minimizing, hence the equality $d(p, \gamma(t)) = t$ holds along γ.

1.3 Sub-Riemannian Manifolds

The distance d of Definition 1.3 does not always meet the classical notion of sub-Riemannian distance arising from a sub-Riemannian manifold. Let us recall the latter definition.

A *sub-Riemannian manifold* (M, D, g_R) is a smooth manifold M endowed with a *sub-Riemannian structure* (D, g_R), where:

- D is a distribution on M, that is a subbundle of TM;
- g_R is a Riemannian metric on D, that is a smooth function $g_R \colon D \to \mathbb{R}$ whose restrictions to $D(q)$ are positive definite quadratic forms.

The *sub-Riemannian metric* associated with (D, g_R) is the function $g_{SR} \colon TM \to \overline{\mathbb{R}}$ given by

$$g_{SR}(q, v) = \begin{cases} g_R(q, v) & \text{if } v \in D(q), \\ +\infty & \text{otherwise.} \end{cases} \tag{1.7}$$

The *sub-Riemannian distance* d_{SR} on M is then defined from the metric g_{SR} as d is defined from the metric g in Definition 1.3.

What is the difference between the two constructions, that is, between the Definitions (1.5) and (1.7) of a sub-Riemannian metric?

Consider a *sub-Riemannian structure* (D, g_R). Locally, on some open subset U, there exist vector fields $X_1, \dots, X_m$ whose values at each point $q \in U$ form an orthonormal basis of $D(q)$ for the quadratic form g_R. The metric g_{SR} associated with (D, g_R) then coincides with the metric g associated with $X_1, \dots, X_m$. Thus, locally, there is a one-to-one correspondence between sub-Riemannian structures and nonholonomic systems for which the dimension of $\Delta(q) = \mathrm{span}\, \{X_1(q), \dots, X_m(q)\}$ is constantly equal to m.

However this correspondence does not hold globally since, for topological reasons, a distribution of rank m may not always be generated by m vector fields on the whole M. Conversely, the vector fields $X_1, \dots, X_m$ of a nonholonomic system do not always generate a linear space $\Delta(q)$ of constant dimension equal to m. It may even be impossible, again for topological reasons (for instance, on an even dimensional sphere).

A way to conciliate both notions is to generalize the definition of sub-Riemannian structure.

Definition 1.4 A *generalized sub-Riemannian structure* on M is a triple (E, σ, g_R) where

- E is a vector bundle over M;
- $\sigma \colon E \to TM$ is a morphism of vector bundles;
- g_R is a Riemannian metric on E.

With a generalized sub-Riemannian structure is associated a metric which is defined by

$$g_{SR}(q, v) = \inf\{g_R(q, u) \colon u \in E(q),\ \sigma(u) = v\}, \quad \text{for } q \in M,\ v \in T_q M.$$

The generalized sub-Riemannian distance d_{SR} on M is then defined from this metric g_{SR} as d is defined from the metric g.

This definition of sub-Riemannian distance actually contains the two notions of distance we have introduced before.

- Take $E = M \times \mathbb{R}^m, \sigma: E \to TM, \sigma(q, u) = \sum_{i=1}^m u_i X_i(q)$ and g_R the Euclidean metric on $\mathbb{R}^m$. The resulting generalized sub-Riemannian distance is the distance associated with the nonholonomic system (1.3).
- Take $E = D$, where D is a distribution on M, $\sigma: D \hookrightarrow TM$ the inclusion, and g_R a Riemannian metric on D. We recover the distance associated with the sub-Riemannian structure (D, g_R).

Locally, a generalized sub-Riemannian structure can always be defined by a single finite family $X_1, \ldots, X_m$ of vector fields, and so by a nonholonomic system (without rank condition). It actually appears that this is also true globally (see [1], or [8] for the fact that a submodule of TM is finitely generated): any generalized sub-Riemannian distance may be associated with a nonholonomic system.

In these notes, we will always consider a sub-Riemannian distance d associated with a nonholonomic system. However, as we just noticed, all the results actually hold for a generalized sub-Riemannian distance.

1.4 Controllability

Given a nonholonomic system,

$$\dot{q} = \sum_{i=1}^m u_i X_i(q), \quad q \in M, \tag{1.8}$$

the first question from control theory is the one of the *controllability*: can we join any two points by a trajectory?

Definition 1.5 The *reachable set from* $p \in M$ is defined to be the set $\mathcal{R}_p$ of points reached by a trajectory of (1.3) issued from p.

The question above then becomes: is the reachable set from any point equal to the whole manifold M? When it is the case the system is said to be *controllable*.

It appears that the controllability of the nonholonomic system (1.3) is mainly characterized by the properties of the Lie algebra generated by $X_1, \ldots, X_m$. We first introduce notions and definitions on this topic.

Let $VF(M)$ denote the set of smooth vector fields on M. We define Δ^1 to be the linear subspace of $VF(M)$ generated by $X_1, \ldots, X_m$,

$$\Delta^1 = \mathrm{span}\{X_1, \ldots, X_m\}.$$

For $s \geq 1$, define $\Delta^{s+1} = \Delta^s + [\Delta^1, \Delta^s]$, where we have set $[\Delta^1, \Delta^s] = \mathrm{span}\{[X, Y]: X \in \Delta^1, Y \in \Delta^s\}$. The *Lie algebra generated by* $X_1, \ldots, X_m$ is defined to be $Lie(X_1, \ldots, X_m) = \bigcup_{s \geq 1} \Delta^s$. Due to the Jacobi identity, $Lie(X_1, \ldots, X_m)$ is the smallest linear subspace of $VF(M)$ which both contains $X_1, \ldots, X_m$ and is invariant by Lie brackets.

Let us denote by $\mathscr{L}(1, \ldots, m)$ the free Lie algebra generated by the elements $\{1, \ldots, m\}$. Recall that $\mathscr{L}(1, \ldots, m)$ is the $\mathbb{R}$-vector space generated by $\{1, \ldots, m\}$ and their formal brackets $[,]$, together with the relations of skew-symmetry and the Jacobi identity enforced. The length of an element I of $\mathscr{L}(1, \ldots, m)$, denoted by $|I|$, is defined inductively by $|I| = 1$ for $I = 1, \ldots, m$, $|I| = |I_1| + |I_2|$ for $I = [I_1, I_2]$. With every element $I \in \mathscr{L}(1, \ldots, m)$ we associate the vector field $X_I \in Lie(X_1, \ldots, X_m)$ obtained by plugging in X_i, $i = 1, \ldots, m$, for the corresponding letter i in I. For instance, $X_{[1,2]} = [X_1, X_2]$. Due to the Jacobi identity, $\Delta^s = \mathrm{span}\{X_I : |I| \leq s\}$.

For $q \in M$, we set $Lie(X_1, \ldots, X_m)(q) = \{X(q) : X \in Lie(X_1, \ldots, X_m)\}$, and, for $s \geq 1$, $\Delta^s(q) = \{X(q) : X \in \Delta^s\}$. By definition these sets are linear subspaces of $T_q M$.

Definition 1.6 We say that (1.3) (or the vector fields $X_1, \ldots, X_m$) satisfies *Chow's Condition* if

$$Lie(X_1, \ldots, X_m)(q) = T_q M, \quad \forall q \in M.$$

Equivalently, for any $q \in M$, there exists an integer $r = r(q)$ such that $\dim \Delta^r(q) = n$.

This property is also known as the Lie algebra rank condition (LARC) in control theory, and as the Hörmander condition in the context of PDEs.

Lemma 1.1 *If* (1.3) *satisfies Chow's Condition, then for every* $p \in M$, *the reachable set* $\mathscr{R}_p$ *is a neighbourhood of* p.

Proof We work in a small neighbourhood $U \subset M$ of p that we identify with a neighbourhood of 0 in $\mathbb{R}^n$.

Let $\phi_t^i = \exp(t X_i)$ be the flow of the vector field X_i, $i = 1, \ldots, m$. Every curve $t \mapsto \phi_t^i(q)$ is a trajectory of (1.3) and we have

$$\phi_t^i = \mathrm{id} + t X_i + o(t).$$

For every element $I \in \mathscr{L}(1, \ldots, m)$, we define the local diffeomorphisms ϕ_t^I on U by induction on the length $|I|$ of I: if $I = [I_1, I_2]$, then

$$\phi_t^I = [\phi_t^{I_1}, \phi_t^{I_2}] := \phi_{-t}^{I_2} \circ \phi_{-t}^{I_1} \circ \phi_t^{I_2} \circ \phi_t^{I_1}.$$

By construction, ϕ_t^I may be expanded as a composition of flows of the vector field X_i, $i = 1, \ldots, m$. As a consequence, $\phi_t^I(q)$ is the endpoint of a trajectory of (1.3) issued from q. Moreover, on a neighbourhood of p there holds

$$\phi_t^I = \mathrm{id} + t^{|I|} X_I + o(t^{|I|}). \tag{1.9}$$

We postpone the proof of this formula to the appendix (Proposition A.1).

To obtain a diffeomorphism whose derivative with respect to the time is exactly X_I, we set

$$
\psi_t^I = \begin{cases}
\phi_{t^{1/|I|}}^I & \text{if } t \geq 0, \\
\phi_{-|t|^{1/|I|}}^I & \text{if } t < 0 \text{ and } |I| \text{ is odd}, \\
[\phi_{|t|^{1/|I|}}^{I_2}, \phi_{|t|^{1/|I|}}^{I_1}] & \text{if } t < 0 \text{ and } |I| \text{ is even},
\end{cases}
$$

where $I = [I_1, I_2]$. Thus

$$
\psi_t^I = \mathrm{id} + t X_I + o(t), \tag{1.10}
$$

and $\psi_t^I(q)$ is the endpoint of a trajectory of (1.3) issued from q.

Let us choose now commutators $X_{I_1}, \dots, X_{I_n}$ whose values at p span $T_p M$. This is possible thanks to Chow's Condition. We introduce the map φ defined on a small neighbourhood Ω of 0 in $\mathbb{R}^n$ by

$$
\varphi(t_1, \dots, t_n) = \psi_{t_n}^{I_n} \circ \cdots \circ \psi_{t_1}^{I_1}(p) \in M.
$$

We conclude from (1.10) that this map is C^1 near 0 and has an invertible derivative at 0, which implies that it is a local C^1-diffeomorphism. Therefore $\varphi(\Omega)$ contains a neighbourhood of p.

Now, for every $t \in \Omega$, $\varphi(t)$ is the endpoint of a concatenation of trajectories of (1.3), the first one being issued from p. It is then the endpoint of a trajectory starting from p. Therefore $\varphi(\Omega) \subset \mathcal{R}_p$, which implies that $\mathcal{R}_p$ is a neighbourhood of p. $\square$

Theorem 1.1 (Chow-Rashevsky's theorem) *If M is connected and if (1.3) satisfies Chow's Condition, then any two points of M can be joined by a trajectory of (1.3).*

Proof Let $p \in M$. If $q \in \mathcal{R}_p$, then $p \in \mathcal{R}_q$. As a consequence, $\mathcal{R}_p = \mathcal{R}_q$ for any $q \in M$ and the lemma above implies that $\mathcal{R}_p$ is an open set. Hence the manifold M is covered by the union of the sets $\mathcal{R}_p$ that are pairwise disjointed. Since M is connected, there is only one such open set. $\square$

Remark 1.3 There exist other proofs of Chow-Rashevsky's theorem, either by using regular controls as in [4], or by imitating the proof of the Krener Theorem as in [1]. Ours is not the more elegant but it has two advantages. First it provides some rough estimates on the sub-Riemannian distance d (see Theorem 1.2 below). Second it is almost constructive, in the sense that the map φ can be used to design a control steering (1.3) from p to $\varphi(t)$. This may lead to solutions to the motion planning problem (see Chap. 3 and more particularly Sect. 3.2.3).

Remark 1.4 This theorem appears also as a consequence of the Orbit Theorem (Stefan [5], Sussmann [6]) since the latter asserts that each set $\mathcal{R}_p$ is a connected immersed submanifold of M whose tangent space $T_q \mathcal{R}_p$ at any point $q \in \mathcal{R}_p$ contains $Lie(X_1, \dots, X_m)(q)$. Note that when the dimension of that Lie algebra is constant on M, we have $Lie(X_1, \dots, X_m)(q) = T_q \mathcal{R}_p$ for every $q \in \mathcal{R}_p$. Thus in this case the vector fields $X_1, \dots, X_m$ restricted to the manifold $\mathcal{R}_p$ always satisfy Chow's Condition.

Remark 1.5 The converse of Chow's theorem is false in general. Consider for instance the nonholonomic system in $\mathbb{R}^3$ defined by $X_1 = \partial_x$, $X_2 = \partial_y + f(x)\partial_z$ where $f(x) = e^{-1/x^2}$ for positive x and $f(x) = 0$ otherwise. The associated sub-Riemannian distance is finite whereas $X_1, \ldots, X_m$ do not satisfy Chow's Condition. However, for an analytic nonholonomic system (i.e. when M and the vector fields $X_1, \ldots, X_m$ are in the analytic category), Chow's Condition is equivalent to the controllability of (1.3) (see [3, 7]).

Remark 1.6 Our proof of Theorem 1.1 also shows that, under the assumptions of the theorem, for every point $p \in M$ the set

$$\left\{\exp(t_{i_1} X_{i_1}) \circ \cdots \circ \exp(t_{i_k} X_{i_k})(p) : k \in \mathbb{N}, \ t_{i_j} \in \mathbb{R}, \ i_j \in \{1, \ldots, m\}\right\}$$

is equal to the whole M. This set is often called the *orbit* at p of the vector fields $X_1, \ldots, X_m$.

The proof of Lemma 1.1 gives a little bit more than the openness of $\mathscr{R}_p$. For ε small enough, any $\phi_t^i(q)$, $0 \leq t \leq \varepsilon$, is a trajectory of length ε. Thus $\varphi(t_1, \ldots, t_n)$ is the endpoint of a trajectory of length less than $N\left(|t_1|^{1/|I_1|} + \cdots + |t_n|^{1/|I_n|}\right)$, where N counts the maximal number of concatenations involved in the $\psi_t^{I_i}$'s. This gives an upper bound for the distance,

$$d\left(p, \varphi(t)\right) \leq N\left(|t_1|^{1/|I_1|} + \cdots + |t_n|^{1/|I_n|}\right). \tag{1.11}$$

This kind of estimates of the distance in terms of local coordinates plays an important role in sub-Riemannian geometry, as we will see in Sect. 2.2.1. However here $(t_1, \ldots, t_n)$ are not smooth local coordinates, as φ is only a C^1-diffeomorphism, not a smooth diffeomorphism.

Let us try to replace $(t_1, \ldots, t_n)$ by smooth local coordinates. Choose local coordinates $(y_1, \ldots, y_n)$ centered at p such that $\frac{\partial}{\partial y_i}|_p = X_{I_i}(p)$. The map $\varphi^y = y \circ \varphi$ is a C^1-diffeomorphism between neighbourhoods of 0 in $\mathbb{R}^n$, and its differential at 0 is $d\varphi_0^y = \mathrm{Id}_{\mathbb{R}^n}$.

Denoting by $\|\cdot\|_{\mathbb{R}^n}$ the Euclidean norm on $\mathbb{R}^n$ we obtain, for $\|t\|_{\mathbb{R}^n}$ small enough, $y_i(t) = t_i + o(\|t\|_{\mathbb{R}^n})$. The inequality (1.11) becomes

$$d(p, q^y) \leq N' \|y\|_{\mathbb{R}^n}^{1/r},$$

where q^y denotes the point of coordinates y, and $r = \max_i |I_i|$. This inequality allows to compare d to a Riemannian distance.

Let g_R be a Riemannian metric on M, and d_R be the associated Riemannian distance. On a compact neighbourhood of p, there exists a constant $c > 0$ such that $g_R(X_i, X_i)(q) \leq c^{-1}$, which implies $c d_R(p, q) \leq d(p, q)$. Moreover we have $d_R(p, q^y) \geq Cst \|y\|_{\mathbb{R}^n}$. We have then obtained a first estimate to the sub-Riemannian distance.

Theorem 1.2 *Assume* (1.3) *satisfies Chow's Condition. For any Riemannian metric* g_R *we have, for q close enough to p,*

$$cd_R(p,q) \leq d(p,q) \leq Cd_R(p,q)^{1/r},$$

where c, C are positive constants and r is an integer such that $\Delta_p^r = T_p M$.

Remark 1.7 If we choose for g_R a Riemannian metric which is compatible with g, that is, which satisfies $g_R|_\Delta = g$, then by construction $d_R(p,q) \leq d(p,q)$.

Corollary 1.1 *Under the hypotheses of Theorem 1.1, d is a distance function on M, i.e.,*

(i) *d is a function from $M \times M$ to $[0, \infty)$;*
(ii) *$d(p,q) = d(q,p)$ (symmetry);*
(iii) *$d(p,q) = 0$ if and only if $p = q$;*
(iv) *$d(p,q) + d(q,q') \leq d(p,q')$ (triangle inequality).*

Proof By Chow-Rashevsky's theorem (Theorem 1.1), the distance between any pair of points is finite, which gives (i). The symmetry of the distance results from the fact that, if $\gamma(s)$, $s \in [0, T]$, is a trajectory joining p to q, then $s \mapsto \gamma(T - s)$ is a trajectory of same length joining q to p. Point (iii) follows directly from Theorem 1.2. Finally, the triangle inequality is a consequence of the following remark. If $\gamma(s)$, $s \in [0, T]$, is a trajectory joining p to q and $\gamma'(s)$, $s \in [0, T']$, is a trajectory joining q to q', then the concatenation $\gamma * \gamma'$, defined by

$$\gamma * \gamma'(s) = \begin{cases} \gamma(s) & \text{if } s \in [0, T], \\ \gamma'(s - T) & \text{if } s \in [T, T + T'], \end{cases}$$

is a trajectory joining p to q' whose length satisfies

$$\text{length}(\gamma * \gamma') = \text{length}(\gamma) + \text{length}(\gamma'). \qquad \square$$

A second consequence of Theorem 1.2 is that the sub-Riemannian distance d is $1/r$-Hölder with respect to any Riemannian distance, and so continuous.

Corollary 1.2 *If* (1.3) *satisfies Chow's Condition, then the topology of the metric space (M, d) coincides with the topology of M as a smooth manifold.*

References

1. Agrachev, A., Barilari, D., Boscain, U.: Introduction to Riemannian and sub-Riemannian geometry (from Hamiltonian viewpoint). Preprint SISSA 09/2012/M (2012)
2. Lee, E.B., Markus, L.: Foundations of Optimal Control Theory. Wiley, New York (1967)
3. Nagano, T.: Linear differential systems with singularities and an application to transitive Lie algebras. J. Math. Soc. Japan **18**, 398–404 (1966)

4. Rifford, L.: Sub-Riemannian Geometry and Optimal Transport (Springer International Publishing), SpringerBriefs in Mathematics (2014)
5. Stefan, P.: Accessible sets, orbits, and foliations with singularities. Proc. London Math. Soc. **29**(3), 699–713 (1974)
6. Sussmann, H.J.: Orbits of families of vector fields and integrability of distributions. Trans. Amer. Math. Soc. **180**, 171–188 (1973)
7. Sussmann, H.J.: An extension of theorem of Nagano on transitive Lie algebras. Proc. Amer. Math. Soc. **45**, 349–356 (1974)
8. Sussmann, H.J.: Smooth distributions are globally finitely spanned. In: Analysis and Design of Nonlinear Control Systems, pp. 3–8. Springer, Berlin (2008)

Chapter 2
First-Order Theory

Abstract Consider a nonholonomic system $\dot{q} = \sum_{i=1}^{m} u_i X_i(q)$, on a manifold M satisfying Chow's Condition, and denote by d the induced sub-Riemannian distance. As we have seen in Sect. 1.1, the infinitesimal behaviour of this system should be captured by an approximation to the first-order with respect to d. In this chapter we will then provide a notion of first-order approximation and construct the basis of an infinitesimal calculus adapted to nonholonomic systems. To this aim, a fundamental role will be played by the concept of nonholonomic order of a function at a point, which is introduced in Sect. 2.1. We will then see that approximations to the first-order appear as nilpotent approximations, in the sense that $X_1, \ldots, X_m$ are approximated by vector fields that generate a nilpotent Lie algebra. In Sect. 2.2 we will use these approximations to obtain estimates of the sub-Riemannian distance in terms of privileged coordinates. We will give a purely metric interpretation of this first-order theory in Sect. 2.3, and show how the distance estimates allow us to compute Hausdorff dimensions. Finally, it will appear along the chapter that singularities of the Lie algebra generated by the vector fields $X_1, \ldots, X_m$ cause qualitative changes in the approximations. We will then show in Sect. 2.4 how to get rid of these singularities by a process of lifting.

Keywords Sub-Riemannian geometry · Nonholonomic systems · Nilpotent systems · Carnot-Carathéodory metric spaces

2.1 First-Order Approximations

The whole section is concerned with local objects. Henceforth, throughout the chapter we fix a point $p \in M$ and an open neighbourhood U of p that we identify, when necessary, with a neighbourhood of 0 in $\mathbb{R}^n$ through some local coordinates.

2.1.1 Nonholonomic Orders

Definition 2.1 Let $f : M \to \mathbb{R}$ be a continuous function. The *nonholonomic order of f at p*, denoted by $\mathrm{ord}_p(f)$, is the real number defined by

© The Author(s) 2014

F. Jean, *Control of Nonholonomic Systems: From Sub-Riemannian Geometry to Motion Planning*, SpringerBriefs in Mathematics, DOI 10.1007/978-3-319-08690-3_2

$$\mathrm{ord}_p(f) = \sup\left\{s \in \mathbb{R} \ : \ f(q) = O\big(d(p,q)^s\big)\right\}.$$

This order is always nonnegative. Moreover $\mathrm{ord}_p(f) = 0$ if $f(p) \neq 0$, and $\mathrm{ord}_p(f) = +\infty$ if $f(p) \equiv 0$.

Example 2.1 (Euclidean case) When $M = \mathbb{R}^n$, $m = n$, and $X_i = \partial_{x_i}$, the sub-Riemannian distance is simply the Euclidean distance on $\mathbb{R}^n$. In this case, non-holonomic orders coincide with the standard ones. Namely, for a smooth function f, $\mathrm{ord}_0(f)$ is the smallest degree of monomials having nonzero coefficient in the Taylor series

$$f(x) \sim \sum c_\alpha x_1^{\alpha_1} \ldots x_n^{\alpha_n}$$

of f at 0. We will see below that there exists in general an analogous characterization of nonholonomic orders.

Let $C^\infty(p)$ denote the set of germs of smooth functions at p. For $f \in C^\infty(p)$, we call *nonholonomic derivatives of order 1 of f* the Lie derivatives $X_1 f, \ldots, X_m f$. We call further $X_i(X_j f)$, $X_i(X_j(X_k f))$,... the *nonholonomic derivatives of f of order* 2, 3,... The nonholonomic derivative of order 0 of f at p is $f(p)$.

Proposition 2.1 *Let $f \in C^\infty(p)$. Then $\mathrm{ord}_p(f)$ is equal to the biggest integer k such that all nonholonomic derivatives of f of order smaller than k vanish at p. Moreover,*

$$f(q) = O\big(d(p,q)^{\mathrm{ord}_p(f)}\big).$$

Proof The proposition results from the following two assertions:

(i) if ℓ is an integer such that $\ell < \mathrm{ord}_p(f)$, then all nonholonomic derivatives of f of order $\leq \ell$ vanish at p;
(ii) if ℓ is an integer such that all nonholonomic derivatives of f of order $\leq \ell$ vanish at p, then $f(q) = O\big(d(p,q)^{\ell+1}\big)$.

Let us first prove point *(i)*. Let ℓ be an integer such that $\ell < \mathrm{ord}_p(f)$. We write a nonholonomic derivative of f of order $k \leq \ell$ as

$$(X_{i_1} \ldots X_{i_k} f)(p) = \frac{\partial^k}{\partial t_1 \ldots \partial t_k} f\big(\exp(t_k X_{i_k}) \circ \cdots \circ \exp(t_1 X_{i_1})(p)\big)\Big|_{t=0}.$$

The point $q = \exp(t_k X_{i_k}) \circ \cdots \circ \exp(t_1 X_{i_1})(p)$ is the endpoint of a trajectory of length $|t_1| + \cdots + |t_n|$. Therefore, $d(p,q) \leq |t_1| + \cdots + |t_n|$.

Since $k \leq \ell < \mathrm{ord}_p(f)$, there exists a real number $s > 0$ such that $f(q) = O\big((|t_1| + \cdots + |t_n|)^{k+s}\big)$. This implies that

$$(X_{i_1} \ldots X_{i_k} f)(p) = \frac{\partial^k}{\partial t_1 \ldots \partial t_k} f(q)\Big|_{t=0} = 0.$$

Thus point (i) is proved.

The proof of point (ii) goes by induction on ℓ. For $\ell = 0$, assume that all non-holonomic derivatives of f of order ≤ 0 vanish at p, that is $f(p) = 0$. Choose any Riemannian metric on M and denote by d_R the associated Riemannian distance on M. Since f is smooth, there holds $f(q) \leq Cst\, d_R(p, q)$ near p. By Theorem 1.2, this implies $f(q) \leq Cst\, d(p, q)$, and so property (ii) for $\ell = 0$.

Assume that, for a given $\ell \geq 0$, (ii) holds for any function f (induction hypothesis) and take a function f such that all its nonholonomic derivatives of order $< \ell + 1$ vanish at p.

Observe that, for $i = 1, \ldots, m$, all the nonholonomic derivatives of $X_i f$ of order $< \ell$ vanish at p. Indeed, $X_{i_1} \ldots X_{i_k}(X_i f) = X_{i_1} \ldots X_{i_k} X_i f$. Applying the induction hypothesis to $X_i f$ leads to $X_i f(q) = O\big(d(p, q)^\ell\big)$. In other words, there exist positive constants $C_1, \ldots, C_m$ such that, for q close enough to p,

$$X_i f(q) \leq C_i d(p, q)^\ell.$$

Fix now a point q near p. By Remark 1.2, there exists a minimizing curve $\gamma(\cdot)$ of velocity one joining p to q. Therefore γ satisfies

$$\dot{\gamma}(t) = \sum_{i=1}^{m} u_i(t) X_i\big(\gamma(t)\big) \quad \text{for a.e. } t \in [0, T], \qquad \gamma(0) = p, \ \gamma(T) = q,$$

with $\sum_i u_i^2(t) = 1$ a.e. and $d\big(p, \gamma(t)\big) = t$ for any $t \in [0, T]$. In particular $d(p, q) = T$.

To estimate $f(q) = f\big(\gamma(T)\big)$, we compute the derivative of $f\big(\gamma(t)\big)$ with respect to t,

$$\frac{d}{dt} f\big(\gamma(t)\big) = \sum_{i=1}^{m} u_i(t) X_i f\big(\gamma(t)\big),$$

$$\Rightarrow \left| \frac{d}{dt} f\big(\gamma(t)\big) \right| \leq \sum_{i=1}^{m} |u_i(t)| C_i d\big(p, \gamma(t)\big)^\ell \leq C t^\ell,$$

where $C = C_1 + \cdots + C_m$. Integrating this inequality between 0 and t gives

$$\left| f\big(\gamma(t)\big) \right| \leq |f(p)| + \frac{C}{\ell + 1} t^{\ell+1}.$$

Note that $f(p) = 0$ since the nonholonomic derivative of f of order 0 at p vanishes. Finally, at $t = T = d(p, q)$ we obtain

$$|f(q)| \leq \frac{C}{\ell + 1} T^{\ell+1},$$

which concludes the proof of (ii). $\qquad\qquad\qquad\qquad\qquad\qquad\qquad\qquad\quad \square$

As a consequence, the nonholonomic order of a smooth (germ of) function is given by the formula

$$\operatorname{ord}_p(f) = \min\left\{s \in \mathbb{N} : \exists\, i_1, \ldots, i_s \in \{1, \ldots, m\} \text{ s.t. } (X_{i_1} \ldots X_{i_s} f)(p) \neq 0\right\},$$

where as usual we adopt the convention that $\min \emptyset = +\infty$.

It is clear now that any function in $C^\infty(p)$ vanishing at p is of order ≥ 1. Moreover, the following basic computation rules are satisfied: for every f, g in $C^\infty(p)$ and every $\lambda \in \mathbb{R} \setminus \{0\}$,

$$\begin{aligned}
\operatorname{ord}_p(fg) &\geq \operatorname{ord}_p(f) + \operatorname{ord}_p(g), \\
\operatorname{ord}_p(\lambda f) &= \operatorname{ord}_p(f), \\
\operatorname{ord}_p(f + g) &\geq \min\left(\operatorname{ord}_p(f), \operatorname{ord}_p(g)\right).
\end{aligned}$$

Note that the first inequality is actually an equality. However the proof of this fact requires an additional result (see Proposition 2.2).

The notion of nonholonomic order extends to vector fields. Let $VF(p)$ denote the set of germs of smooth vector fields at p.

Definition 2.2 Let $X \in VF(p)$. The *nonholonomic order of X at p*, denoted by $\operatorname{ord}_p(X)$, is the real number defined by:

$$\operatorname{ord}_p(X) = \sup\left\{\sigma \in \mathbb{R} : \operatorname{ord}_p(Xf) \geq \sigma + \operatorname{ord}_p(f), \quad \forall f \in C^\infty(p)\right\}.$$

The order of a differential operator is defined in the same way.

Note that $\operatorname{ord}_p(X) \in \mathbb{Z}$ since the order of a smooth function is an integer. Moreover the null vector field $X \equiv 0$ has infinite order, $\operatorname{ord}_p(0) = +\infty$.

Since the order of a function coincides with its order as a differential operator acting by multiplication, we have the following properties. For every $X, Y \in VF(p)$ and every $f \in C^\infty(p)$,

$$\begin{aligned}
\operatorname{ord}_p([X, Y]) &\geq \operatorname{ord}_p(X) + \operatorname{ord}_p(Y), \\
\operatorname{ord}_p(fX) &\geq \operatorname{ord}_p(f) + \operatorname{ord}_p(X), \\
\operatorname{ord}_p(X) &\leq \operatorname{ord}_p(Xf) - \operatorname{ord}_p(f), \\
\operatorname{ord}_p(X + Y) &\geq \min\left(\operatorname{ord}_p(X), \operatorname{ord}_p(Y)\right).
\end{aligned} \tag{2.1}$$

As already noticed for functions, the second inequality is in fact an equality. This is not the case for the first inequality (take for instance $X = Y$). As a consequence of (2.1), $X_1, \ldots, X_m$ are of order ≥ -1, $[X_i, X_j]$ of order ≥ -2, and more generally, every X in the set Δ^k is of order $\geq -k$.

Example 2.2 (Euclidean case) In the Euclidean case (see Example 2.1), the nonholonomic order of a constant differential operator is the negative of its usual order. For instance ∂_{x_i} is of nonholonomic order -1. Actually, in this case, every vector field that does not vanish at p is of nonholonomic order -1.

Example 2.3 (Heisenberg case) Consider the following vector fields on $\mathbb{R}^3$:

$$X_1 = \partial_x - \frac{y}{2}\partial_z \quad \text{and} \quad X_2 = \partial_y + \frac{x}{2}\partial_z.$$

The coordinate functions x and y have order 1 at 0, whereas z has order 2 at 0, since $X_1 x(0) = X_2 y(0) = 1$, $X_1 z(0) = X_2 z(0) = 0$, and $X_1 X_2 z(0) = 1/2$. These relations also imply $\mathrm{ord}_0(X_1) = \mathrm{ord}_0(X_2) = -1$. Finally, the Lie bracket $[X_1, X_2] = \partial_z$ is of order -2 at 0 since $[X_1, X_2]z = 1$.

We are now in a position to give a meaning to first-order approximations.

Definition 2.3 A family of m vector fields $(\widehat{X}_1, \ldots, \widehat{X}_m)$ defined near p is called a *first-order approximation of* $(X_1, \ldots, X_m)$ *at* p if the vector fields $X_i - \widehat{X}_i$, $i = 1, \ldots, m$, are of order ≥ 0 at p.

A consequence of this definition is that the order at p defined by the vector fields $\widehat{X}_1, \ldots, \widehat{X}_m$ coincides with the one defined by $X_1, \ldots, X_m$. Hence for any $f \in C^\infty(p)$ of order greater than $s - 1$,

$$(X_{i_1} \ldots X_{i_s} f)(q) = (\widehat{X}_{i_1} \ldots \widehat{X}_{i_s} f)(q) + O\left(d(p,q)^{\mathrm{ord}_p(f)-s+1}\right).$$

To go further in the characterization of orders and approximations, we need suitable systems of coordinates.

2.1.2 Privileged Coordinates

We have introduced in Sect. 1.4 the sets of vector fields Δ^s, defined by $\Delta^s = \mathrm{span}\{X_I : |I| \leq s\}$. Since $X_1, \ldots, X_m$ satisfy Chow's Condition, the values of these sets at p form a flag of subspaces of $T_p M$, that is,

$$\Delta^1(p) \subset \Delta^2(p) \subset \cdots \subset \Delta^{r-1}(p) \subsetneq \Delta^r(p) = T_p M, \tag{2.2}$$

where $r = r(p)$ is called the *degree of nonholonomy at* p.

Set $n_i(p) = \dim \Delta^i(p)$. The r-tuple of integers $(n_1(p), \ldots, n_r(p))$ is called the *growth vector at* p. The first integer in the growth vector is the rank $n_1(p) \leq m$ of the family $X_1(p), \ldots, X_m(p)$, and the last one $n_r(p) = n$ is the dimension of the manifold M.

Let $s \geq 1$. By abuse of notations, we continue to write Δ^s for the map $q \mapsto \Delta^s(q)$. This map Δ^s is a distribution if and only if $n_s(q)$ is constant on M. We then distinguish two kind of points.

Definition 2.4 The point p is a *regular point* (w.r.t. $X_1, \ldots, X_m$) if the growth vector is constant in a neighbourhood of p. Otherwise, p is a *singular point*.

Thus, near a regular point, all maps Δ^s are locally distributions.

The structure of the flag (2.2) may also be described by another sequence of integers. We define the *weights at* p, $w_i = w_i(p)$, $i = 1, \dots, n$, by setting $w_j = s$ if $n_{s-1}(p) < j \leq n_s(p)$, where $n_0 = 0$. In other words, we have

$$w_1 = \cdots = w_{n_1} = 1, \; w_{n_1+1} = \cdots = w_{n_2} = 2, \; \dots, w_{n_{r-1}+1} = \cdots = w_{n_r} = r.$$

The weights at p form an increasing sequence $w_1(p) \leq \cdots \leq w_n(p)$ which is constant near p if and only if p is a regular point.

Example 2.4 (Heisenberg case) The Heisenberg case in $\mathbb{R}^3$ given in Example 2.3 has a growth vector which is equal to $(2, 3)$ at every point. Therefore all points of $\mathbb{R}^3$ are regular. The weights at any point are $w_1 = w_2 = 1$, $w_3 = 2$.

Example 2.5 (Martinet case) Consider the following vector fields on $\mathbb{R}^3$,

$$X_1 = \partial_x \quad \text{and} \quad X_2 = \partial_y + \frac{x^2}{2}\partial_z.$$

The only nonzero brackets are

$$[X_1, X_2] = x\partial_z \quad \text{and} \quad [X_1, [X_1, X_2]] = \partial_z.$$

Thus the growth vector is equal to $(2, 2, 3)$ on the plane $\{x = 0\}$, and to $(2, 3)$ elsewhere. As a consequence, the set of singular points is the plane $\{x = 0\}$. The weights are $w_1 = w_2 = 1$, $w_3 = 2$ at regular points, and $w_1 = w_2 = 1$, $w_3 = 3$ at singular ones.

Example 2.6 Consider the vector fields on $\mathbb{R}^3$

$$X_1 = \partial_x \quad \text{and} \quad X_2 = \partial_y + f(x)\partial_z,$$

where f is a smooth function on $\mathbb{R}$ which admits every positive integer $n \in \mathbb{N}$ as a zero with multiplicity n (such a function exists and can even be chosen in the analytic class thanks to the Weierstrass factorization theorem [23, Theorem 15.9]). Every point (n, y, z) is singular and the weights at this point are $w_1 = w_2 = 1$, $w_3 = n + 1$. As a consequence the degree of nonholonomy w_3 is unbounded on $\mathbb{R}^3$.

Let us give some basic properties of the growth vector and of the weights.

- At a regular point, the growth vector is a strictly increasing sequence: $n_1(p) < \cdots < n_r(p)$. Indeed, if $n_s(q) = n_{s+1}(q)$ in a neighbourhood of p, then Δ^s is locally an involutive distribution and so $s = r$. As a consequence, at a regular point p, the jump between two successive weights is never greater than 1, $w_{i+1} - w_i \leq 1$, and there holds $r(p) \leq n - m + 1$.
- For every s, the map $q \mapsto n_s(q)$ is a lower semi-continuous function from M to $\mathbb{N}$. Similarly, for every $i = 1, \dots, n$, the weight $w_i(\cdot)$ is an upper semi-continuous

function. This is in particular the case for the degree of nonholonomy $r(\cdot) = w_n(\cdot)$, that is, $r(q) \leq r(p)$ for q near p. As a consequence $r(\cdot)$ is bounded on any compact subset of M.

- The set of regular points is open and dense in M. The openness results from the very definition of regular points. As for the density, take any point p in M and consider an open neighbourhood V of p small enough so that $r(q) \leq r(p)$ for $q \in V$. For every s, the set of point V_s where n_s reaches its maximal value on V is an open and nonempty subset of V due to the lower semi-continuity of n_s. The intersection of all V_s for $s \leq r(p)$ is an open subset of V and contains only regular points since the growth vector is constant on this subset. The density of the set of regular points follows.

- The degree of nonholonomy may be unbounded on M (see Example 2.6 above). Thus determining if a nonholonomic system is controllable is a non decidable problem: the computation of an infinite number of brackets may be needed to decide if Chow's Condition is satisfied. However, in the case of polynomial vector fields on $\mathbb{R}^n$ (relevant in practice), it can be shown that the degree of nonholonomy is bounded by a universal function of the degree k of the polynomials (see [8, 14]):

$$r(x) \leq 2^{3n^2} n^{2n} k^{2n}.$$

The meaning of the sequence of weights is best understood in terms of basis of $T_p M$. Choose first vector fields $Y_1, \ldots, Y_{n_1}$ in Δ^1 whose values at p form a basis of $\Delta^1(p)$. Choose then vector fields $Y_{n_1+1}, \ldots, Y_{n_2}$ in Δ^2 such that the values $Y_1(p), \ldots, Y_{n_2}(p)$ form a basis of $\Delta^2(p)$. For each s, choose $Y_{n_{s-1}+1}, \ldots, Y_{n_s}$ in Δ^s such that $Y_1(p), \ldots, Y_{n_s}(p)$ form a basis of $\Delta^s(p)$. We obtain in this way a family of vector fields $Y_1, \ldots, Y_n$ such that

$$\begin{cases} Y_1(p), \ldots, Y_n(p) \text{ is a basis of } T_p M, \\ Y_i \in \Delta^{w_i}, \; i = 1, \ldots, n. \end{cases} \tag{2.3}$$

A family of n vector fields satisfying (2.3) is called an *adapted frame at p*. The word "adapted" means here "adapted to the flag (2.2)", since the values at p of an adapted frame contain a basis $Y_1(p), \ldots, Y_{n_s}(p)$ of each subspace $\Delta^s(p)$ of the flag. By continuity, at a point q close enough to p the values of $Y_1, \ldots, Y_n$ still form a basis of $T_q M$. However, if p is singular this basis may not be adapted to the flag (2.2) at q.

Let us explain now the relation between weights and orders. We write first the tangent space as a direct sum,

$$T_p M = \Delta^1(p) \oplus \Delta^2(p)/\Delta^1(p) \oplus \cdots \oplus \Delta^r(p)/\Delta^{r-1}(p),$$

where $\Delta^s(p)/\Delta^{s-1}(p)$ denotes a supplementary of $\Delta^{s-1}(p)$ in $\Delta^s(p)$. Let us choose local coordinates $(y_1, \ldots, y_n)$. The dimension of each space $\Delta^s(p)/\Delta^{s-1}(p)$ is equal to $n_s - n_{s-1}$, so we can assume that, up to a reordering, the local coordinates satisfy $dy_j(\Delta^s(p)/\Delta^{s-1}(p)) \neq 0$ for $n_{s-1} < j \leq n_s$.

Take an integer j such that $0 < j \leq n_1$. From the assumption above, there holds $dy_j(\Delta^1(p)) \neq 0$, and consequently there exists X_i such that $dy_j(X_i(p)) \neq 0$. Since $dy_j(X_i) = X_i y_j$ is a first-order nonholonomic derivative of y_j, we have $\mathrm{ord}_p(y_j) \leq 1 = w_j$.

Take now an integer j such that $n_{s-1} < j \leq n_s$ for $s > 1$, that is, $w_j = s$. Since $dy_j(\Delta^s(p)/\Delta^{s-1}(p)) \neq 0$, there exists a vector field Y in Δ^s such that $dy_j(Y(p)) = (Yy_j)(p) \neq 0$. By definition of Δ^s, the Lie derivative Yy_j is a linear combination of nonholonomic derivatives of y_j of order not greater than s. One of them must be nonzero, and so $\mathrm{ord}_p(y_j) \leq s = w_j$.

Finally, any system of local coordinates $(y_1, \ldots, y_n)$ satisfies $\mathrm{ord}_p(y_j) \leq w_j$ up to a reordering (or $\sum_{i=1}^n \mathrm{ord}_p(y_i) \leq \sum_{i=1}^n w_i$ without reordering). The coordinates with the maximal possible order will play an important role.

Definition 2.5 A *system of privileged coordinates at p* is a system of local coordinates $(z_1, \ldots, z_n)$ such that $\mathrm{ord}_p(z_j) = w_j$ for $j = 1, \ldots, n$.

Notice that privileged coordinates $(z_1, \ldots, z_n)$ satisfy

$$dz_i(\Delta^{w_i}(p)) \neq 0, \quad dz_i(\Delta^{w_i-1}(p)) = 0, \quad i = 1, \ldots, n, \tag{2.4}$$

or, equivalently, $\partial_{z_i}|_p$ belongs to $\Delta^{w_i}(p)$ but not to $\Delta^{w_i-1}(p)$. Local coordinates satisfying (2.4) are called *linearly adapted coordinates* ("adapted" because the differentials at p of the coordinates form a basis of $T_p^* M$ dual to the values of an adapted frame). Thus privileged coordinates are always linearly adapted coordinates. The converse is false, as shown in the example below.

Example 2.7 Take $X_1 = \partial_x$, $X_2 = \partial_y + (x^2 + y)\partial_z$ in $\mathbb{R}^3$. The weights at 0 are $(1, 1, 3)$ and (x, y, z) are adapted at 0. But they are not privileged: indeed, the coordinate z is of order 2 at 0 since $(X_2 X_2 z)(0) = 1$.

Remark 2.1 Let us define *privileged functions at p* to be smooth functions f on U such that
$$\mathrm{ord}_p(f) = \max\{s \in \mathbb{N} \ : \ df(\Delta^s(p)/\Delta^{s-1}(p)) \neq 0\},$$

(a similar notion has been suggested by Kupka [17]). It results from the discussion above that some local coordinates $(z_1, \ldots, z_n)$ are privileged at p if and only if each z_i is a privileged function at p.

Let us now show how to compute orders using privileged coordinates. We fix a system of privileged coordinates $(z_1, \ldots, z_n)$ at p. Given a sequence of integers $\alpha = (\alpha_1, \ldots, \alpha_n)$, we define the weighted degree of the monomial $z^\alpha = z_1^{\alpha_1} \ldots z_n^{\alpha_n}$ to be $w(\alpha) = w_1\alpha_1 + \cdots + w_n\alpha_n$ and the weighted degree of the monomial vector field $z^\alpha \partial_{z_j}$ to be $w(\alpha) - w_j$. The weighted degrees allow to compute the orders of functions and vector fields in a purely algebraic way.

Proposition 2.2 *For a smooth function f with a Taylor expansion*

$$f(z) \sim \sum_{\alpha} c_\alpha z^\alpha,$$

the order of f is the least weighted degree of monomials having a nonzero coefficient in the Taylor series.

For a vector field X with a Taylor expansion

$$X(z) \sim \sum_{\alpha, j} a_{\alpha, j} z^\alpha \partial_{z_j},$$

the order of X is the least weighted degree of a monomial vector fields having a nonzero coefficient in the Taylor series.

In other words, when using privileged coordinates, the notion of nonholonomic order amounts to the usual notion of vanishing order at some point, only assigning weights to the variables.

Proof For $i = 1, \ldots, n$, we have $\partial_{z_i}|_p \in \Delta^{w_i}(p)$. Then there exist n vector fields $Y_1, \ldots, Y_n$ which form an adapted frame at p and such that $Y_1(p) = \partial_{z_1}|_p$, ..., $Y_n(p) = \partial_{z_n}|_p$. For every i, the vector field Y_i is of order $\geq -w_i$ at p since it belongs to Δ^{w_i}. Moreover we have $(Y_i z_i)(p) = 1$ and $\mathrm{ord}_p(z_i) = w_i$. Thus $\mathrm{ord}_p(Y_i) = -w_i$.

Take a sequence of integers $\alpha = (\alpha_1, \ldots, \alpha_n)$. The monomial z^α is of order $\geq w(\alpha)$ at p and the differential operator $Y^\alpha = Y_1^{\alpha_1} \ldots Y_n^{\alpha_n}$ is of order $\geq -w(\alpha)$. Observing that $(Y_i z_j)(p) = 0$ if $j \neq i$, we easily see that $(Y^\alpha z^\alpha)(p) = \frac{1}{\alpha_1! \ldots \alpha_n!} \neq 0$, whence $\mathrm{ord}_p(z^\alpha) = w(\alpha)$.

In the same way, we obtain that, if z^α, z^β are two different monomials and λ, μ two nonzero real numbers, then $\mathrm{ord}_p(\lambda z^\alpha + \mu z^\beta) = \min\big(w(\alpha), w(\beta)\big)$. Thus the order of a series is the least weighted degree of monomials actually appearing in the series itself. This shows the statement on order of functions.

As a consequence, for any smooth function f, the order at p of $\partial_{z_i} f$ is $\geq \mathrm{ord}_p(f) - w_i$. Since moreover $\partial_{z_i} z_i = 1$, we obtain that $\mathrm{ord}_p(\partial_{z_i})$ is equal to $-w_i$. The second part of the statement follows. $\qquad\square$

A notion of homogeneity is also naturally associated with a system of privileged coordinates $(z_1, \ldots, z_n)$ defined on an open neighbourhood U of the point p. We define first the one-parameter family of dilations

$$\delta_t : (z_1, \ldots, z_n) \mapsto (t^{w_1} z_1, \ldots, t^{w_n} z_n), \qquad t \geq 0.$$

Each dilation δ_t is a map from $\mathbb{R}^n$ to $\mathbb{R}^n$. By abuse of notations, for $q \in U$ and t small enough we write $\delta_t(q)$ instead of $\delta_t(z(q))$, where $z(q)$ are the coordinates of q. A dilation δ_t acts also on functions and vector fields by pull-back: $\delta_t^* f = f \circ \delta_t$ and $\delta_t^* X$ is the vector field such that $(\delta_t^* X)(\delta_t^* f) = \delta_t^*(Xf)$.

Definition 2.6 A function f is *homogeneous of degree s* if $\delta_t^* f = t^s f$. A vector field X is *homogeneous of degree σ* if $\delta_t^* X = t^\sigma X$.

For a smooth function (resp. a smooth vector field), this is the same as being a finite sum of monomials (resp. monomial vector fields) of weighted degree s. As a consequence, if a function f is homogeneous of degree s, then it is of order s at p.

A typical degree 1 homogeneous function is the so-called *pseudo-norm at p*, defined by:

$$z \mapsto \|z\|_p = |z_1|^{1/w_1} + \cdots + |z_n|^{1/w_n}. \tag{2.5}$$

When composed with the coordinates function, the pseudo-norm at p is a (non smooth) function of order 1, that is,

$$\|z(q)\|_p = O\big(d(p,q)\big).$$

Actually, it results from Proposition 2.2 that the order of a function $f \in C^\infty(p)$ is the least integer s such that $f(q) = O(\|z(q)\|_p^s)$.

Examples of Privileged Coordinates

Of course all the results above on algebraic computation of orders hold only if privileged coordinates do exist. Two types of privileged coordinates are commonly used in the literature.

a. Exponential coordinates

Choose an adapted frame $Y_1, \ldots, Y_n$ at p. The inverse of the local diffeomorphism

$$(z_1, \ldots, z_n) \mapsto \exp(z_1 Y_1 + \cdots + z_n Y_n)(p)$$

defines a system of local privileged coordinates at p, called *canonical coordinates of the first kind*. These coordinates are mainly used in the context of hypoelliptic operator and for nilpotent Lie groups with right (or left) invariant sub-Riemannian structure.

The inverse of the local diffeomorphism

$$(z_1, \ldots, z_n) \mapsto \exp(z_n Y_n) \circ \cdots \circ \exp(z_1 Y_1)(p)$$

also defines privileged coordinates at p, called *canonical coordinates of the second kind*. They are easier to work with than the one of the first kind. For instance, in these coordinates, the vector field Y_n read as ∂_{z_n}. One can also exchange the order of the flows in the definition to obtain any of the Y_i as ∂_{z_i}. The fact that canonical coordinates of first and second kind are privileged is proved in Appendix (Sects. B.1 and B.2).

We leave it to the reader to verify that the diffeomorphism

$$(z_1, \ldots, z_n) \mapsto \exp(z_n Y_n + \cdots + z_{s+1} Y_{s+1}) \circ \exp(z_s Y_s) \cdots \circ \exp(z_1 Y_1)(p)$$

also induces privileged coordinates. As a matter of fact, any "mix" between first and second kind canonical coordinates defines privileged coordinates.

b. Algebraic coordinates

There exist also effective constructions of privileged coordinates (the construction of exponential coordinates is not effective in general since it requires to integrate flows of vector fields). We present here Bellaïche's algorithm, but other constructions exist (see [2, 24]).

1. Choose an adapted frame $Y_1, \ldots, Y_n$ at p.
2. Choose coordinates $(y_1, \ldots, y_n)$ centered at p such that $\partial_{y_i}|_p = Y_i(p)$.
3. For $j = 1, \ldots, n$, set

$$z_j = y_j - \sum_{k=2}^{w_j - 1} h_k(y_1, \ldots, y_{j-1}),$$

where, for $k = 2, \ldots, w_j - 1$,

$$h_k(y_1, \ldots, y_{j-1}) = \sum_{\substack{|\alpha|=k \\ w(\alpha)<w_j}} Y_1^{\alpha_1} \cdots Y_{j-1}^{\alpha_{j-1}} \left(y_j - \sum_{q=2}^{k-1} h_q(y) \right)(p) \frac{y_1^{\alpha_1}}{\alpha_1!} \cdots \frac{y_{j-1}^{\alpha_{j-1}}}{\alpha_{j-1}!},$$

with $|\alpha| = \alpha_1 + \cdots + \alpha_n$.

The fact that coordinates $(z_1, \ldots, z_n)$ are privileged at p will be proved in Sect. B.3.

Coordinates $(y_1, \ldots, y_n)$ are linearly adapted coordinates. They can be obtained from any original system of coordinates by an affine change. The privileged coordinates $(z_1, \ldots, z_n)$ are then obtained from $(y_1, \ldots, y_n)$ by an expression of the form

$$\begin{aligned}
z_1 &= y_1, \\
z_2 &= y_2 + \mathrm{pol}(y_1), \\
&\;\vdots \\
z_n &= y_n + \mathrm{pol}(y_1, \ldots, y_{n-1}),
\end{aligned}$$

where each pol is a polynomial function without constant nor linear terms. The inverse change of coordinates takes the same triangular form, which makes the use of these coordinates easy for computations.

2.1.3 Nilpotent Approximation

Fix a system of privileged coordinates $(z_1, \ldots, z_n)$ at p. Every vector field X_i is of order ≥ -1, hence it has, in z coordinates, a Taylor expansion

$$X_i(z) \sim \sum_{\alpha, j} a_{\alpha, j} z^\alpha \partial_{z_j},$$

where $w(\alpha) \geq w_j - 1$ if $a_{\alpha, j} \neq 0$. Grouping together the monomial vector fields of same weighted degree we express X_i as a series

$$X_i = X_i^{(-1)} + X_i^{(0)} + X_i^{(1)} + \cdots,$$

where $X_i^{(s)}$ is a homogeneous vector field of degree s.

Proposition 2.3 *Set $\widehat{X}_i = X_i^{(-1)}$, $i = 1, \ldots, m$. The family of vector fields $(\widehat{X}_1, \ldots, \widehat{X}_m)$ is a first-order approximation of $(X_1, \ldots, X_m)$ at p and generates a nilpotent Lie algebra of step $r = w_n$.*

Recall that a Lie algebra $Lie(X_1, \ldots, X_m)$ is said to be *nilpotent of step s* if all brackets X_I of length $|I|$ greater than s are zero.

Proof The fact that the vector fields $\widehat{X}_1, \ldots, \widehat{X}_m$ form a first-order approximation of $X_1, \ldots, X_m$ results from their construction. Note further that any homogeneous vector field of degree smaller than $-w_n$ is zero, as it is easy to check in privileged coordinates. Moreover, if X and Y are homogeneous of degree respectively k and l, then the bracket $[X, Y]$ is homogeneous of degree $k + l$ because $\delta_t^*[X, Y] = [\delta_t^* X, \delta_t^* Y] = t^{k+l}[X, Y]$.

It follows that every iterated bracket of the vector fields $\widehat{X}_1, \ldots, \widehat{X}_m$ of length k is homogeneous of degree $-k$ and is zero if $k > w_n$. $\qquad\square$

Definition 2.7 The family $(\widehat{X}_1, \ldots, \widehat{X}_m)$ is called the *(homogeneous) nilpotent approximation* of $(X_1, \ldots, X_m)$ at p associated with the coordinates z.

Example 2.8 (unicycle) Consider the vector fields on $\mathbb{R}^2 \times \mathscr{S}^1$ defining the kinematic model of a unicycle (see Example 1.1), that is, $X_1 = \cos\theta\, \partial_x + \sin\theta\, \partial_y$, $X_2 = \partial_\theta$. We have $[X_1, X_2] = \sin\theta\, \partial_x - \cos\theta\, \partial_y$, so the weights are $(1, 1, 2)$ at every point. At $p = 0$, the coordinates (x, θ) have order 1 and y has order 2, consequently (x, θ, y) is a system of privileged coordinates at 0. Taking the Taylor expansion of X_1 and X_2 in the latter coordinates, we obtain the homogeneous components:

$$X_1^{(-1)} = \partial_x + \theta\, \partial_y, \quad X_1^{(0)} = 0, \quad X_1^{(1)} = -\frac{\theta^2}{2}\partial_x - \frac{\theta^3}{3!}\partial_y, \quad \cdots$$

and $X_2^{(-1)} = X_2 = \partial_\theta$. Therefore the homogeneous nilpotent approximation of (X_1, X_2) at 0 in coordinates (x, θ, y) is

$$\widehat{X}_1 = \partial_x + \theta \partial_y, \quad \widehat{X}_2 = \partial_\theta.$$

We easily check that the Lie brackets of length 3 of these vectors are zero, that is, $[\widehat{X}_1, [\widehat{X}_1, \widehat{X}_2]] = [\widehat{X}_2, [\widehat{X}_1, \widehat{X}_2]] = 0$, and so the Lie algebra $Lie(\widehat{X}_1, \widehat{X}_2)$ is nilpotent of step 2.

The homogeneous nilpotent approximation is not uniquely defined by the m-tuple $(X_1, \ldots, X_m)$, since it depends on the chosen system of privileged coordinates. However, if $\widehat{X}_1, \ldots, \widehat{X}_m$ and $\widehat{X}'_1, \ldots, \widehat{X}'_m$ are the nilpotent approximations associated with two different systems of coordinates, then their Lie algebras $Lie(\widehat{X}_1, \ldots, \widehat{X}_m)$ and $Lie(\widehat{X}'_1, \ldots, \widehat{X}'_m)$ are isomorphic. If moreover p is a regular point, then $Lie(\widehat{X}_1, \ldots, \widehat{X}_m)$ is isomorphic to the graded nilpotent Lie algebra

$$\mathrm{Gr}(\Delta)_p = \Delta(p) \oplus (\Delta^2/\Delta^1)(p) \oplus \cdots \oplus (\Delta^{r-1}/\Delta^r)(p).$$

Remark 2.2 The nilpotent approximation denotes in fact two different objects. Each $\widehat{X}_i$ may be seen as a vector field on $\mathbb{R}^n$ or as the representation in z coordinates of the vector field $z^*\widehat{X}_i$ defined on a neighbourhood of p in M (recall that $z^*\widehat{X}_i$ denotes the pullback $d(z^{-1}) \circ \widehat{X}_i \circ z$ of $\widehat{X}_i$ by the local diffeomorphism z). This will cause no confusion since the nilpotent approximation is associated with a given system of privileged coordinates.

It is worth to notice the particular form of the nilpotent approximation in privileged coordinates. Write $\widehat{X}_i = \sum_{j=1}^n f_{ij}(z)\partial_{z_j}, i = 1, \ldots, m$. Since $\widehat{X}_i$ is homogeneous of degree -1 and ∂_{z_j} of degree $-w_j$, the function f_{ij} is a homogeneous polynomial of weighted degree $w_j - 1$. In particular it can not involve variables of weight greater than $w_j - 1$, that is,

$$\widehat{X}_i(z) = \sum_{j=1}^n f_{ij}(z_1, \ldots, z_{n_{w_j}-1})\partial_{z_j}.$$

The nonholonomic system $\dot{z} = \sum_{i=1}^m u_i \widehat{X}_i(z)$ associated with the nilpotent approximation is then polynomial and in a triangular form,

$$\dot{z}_j = \sum_{i=1}^m u_i f_{ij}(z_1, \ldots, z_{n_{w_j}-1}). \tag{2.6}$$

Computing the trajectories of a system in such a form is rather easy: given the input function $(u_1(t), \ldots, u_m(t))$, it is possible to compute the coordinates z_j one after the other, only by integration. As a consequence, for every input function the associated differential Eq. (2.6) is complete and every $\widehat{X}_i$ is a complete vector field on $\mathbb{R}^n$.

As vector fields on $\mathbb{R}^n$, $\widehat{X}_1, \ldots, \widehat{X}_m$ generate a sub-Riemannian distance on $\mathbb{R}^n$ which is homogeneous with respect to the dilation δ_t.

Lemma 2.1 (i) *The family* $(\widehat{X}_1, \ldots, \widehat{X}_m)$ *satisfies Chow's Condition on* $\mathbb{R}^n$.

(ii) *The growth vector at* 0 *of* $(\widehat{X}_1, \ldots, \widehat{X}_m)$ *is equal to the one at* p *of* $(X_1, \ldots, X_m)$.

(iii) *Let* $\widehat{d}$ *be the sub-Riemannian distance on* $\mathbb{R}^n$ *associated with* $(\widehat{X}_1, \ldots, \widehat{X}_m)$. *The distance* $\widehat{d}$ *is homogeneous of degree* 1,

$$\widehat{d}(\delta_t x, \delta_t y) = t\widehat{d}(x, y).$$

(iv) *There exists a constant* $C > 0$ *such that, for all* $z \in \mathbb{R}^n$,

$$\frac{1}{C}\|z\|_p \leq \widehat{d}(0, z) \leq C\|z\|_p,$$

where the pseudo-norm $\|\cdot\|_p$ *at* p *is defined by* (2.5).

Proof Through the coordinates z we identify the neighbourhood U of p in M with a neighbourhood of 0 in $\mathbb{R}^n$.

For every element $I \in \mathscr{L}(1, \ldots, m)$, we denote by $\widehat{X}_I \in Lie(\widehat{X}_1, \ldots, \widehat{X}_m)$ the vector field that is obtained by plugging in $\widehat{X}_i$, $i = 1, \ldots, m$, for the corresponding letter i in I. For $k \geq 1$ we set $\widehat{\Delta}^k = \mathrm{span}\{\widehat{X}_I \; : \; |I| \leq k\}$. As noticed in the proof of Proposition 2.3, a bracket $\widehat{X}_I$ of length $|I| = k$ is homogeneous of weighted degree $-k$, and by construction of the nilpotent approximation, there holds $X_I = \widehat{X}_I +$ terms of order $> -k$. Therefore,

$$\widehat{X}_I(0) = X_I(p)\,\mathrm{mod\,span}\{\partial_{z_j}\big|_p \; : \; w_j < k\} = X_I(p)\,\mathrm{mod}\,\Delta^{k-1}(p).$$

As a consequence, for any integer $k \geq 1$, we have

$$\dim \widehat{\Delta}^k(0) = \dim \Delta^k(p), \qquad (2.7)$$

and property (ii) follows. Moreover, let us choose an adapted frame $X_{I_1}, \ldots, X_{I_n}$ at p. The family $(\widehat{X}_{I_1}(0), \ldots, \widehat{X}_{I_n}(0))$ is of rank n, which implies that its determinant is nonzero. Since the determinant of $\widehat{X}_{I_1}, \ldots, \widehat{X}_{I_n}$ is an homogeneous polynomial of weighted degree 0, and so a constant, it is nonzero everywhere. This implies (i).

As for the property (iii), consider the nonholonomic system defined by the nilpotent approximation, that is, $\dot{z} = \sum_{i=1}^m u_i \widehat{X}_i(z)$. Observe that, if $\widehat{\gamma}$ is a trajectory of this system, that is, if

$$\dot{\widehat{\gamma}}(t) = \sum_{i=1}^m u_i \widehat{X}_i\big(\widehat{\gamma}(t)\big), \quad t \in [0, T],$$

then the dilated curve $\delta_\lambda \widehat{\gamma}$ satisfies

$$\frac{d}{dt}\delta_\lambda \widehat{\gamma}(t) = \sum_{i=1}^m \lambda u_i \widehat{X}_i\big(\delta_\lambda \widehat{\gamma}(t)\big), \quad t \in [0, T].$$

Thus $\delta_\lambda \widehat{\gamma}$ is a trajectory of the same system, with extremities $(\delta_\lambda \widehat{\gamma})(0) = \delta_\lambda(\widehat{\gamma}(0))$ and $(\delta_\lambda \widehat{\gamma})(T) = \delta_\lambda(\widehat{\gamma}(T))$, and its length equals $\lambda \mathrm{length}(\widehat{\gamma})$. This proves the homogeneity of $\widehat{d}$.

Finally, since $(\widehat{X}_1, \ldots, \widehat{X}_m)$ satisfies Chow's Condition, the distance $\widehat{d}(0, \cdot)$ is continuous on $\mathbb{R}^n$ by Corollary 1.2. We can then choose a real number $C > 0$ such that, on the compact set $\{\|z\| = 1\}$, we have $1/C \leq \widehat{d}(0, z) \leq C$. Both functions $\widehat{d}(0, z)$ and $\|z\|$ being homogeneous of degree 1, the inequality of Property (iv) follows. $\qquad\square$

2.2 Distance Estimates

2.2.1 Local Distance Estimates

As it is the case for Riemannian distances, in general it is impossible to compute analytically a sub-Riemannian distance (it would require to determine all minimizing curves). Therefore it is very important to obtain estimates of the distance, at least locally. In a Riemannian manifold (M, g), the situation is rather simple: in local coordinates x centered at a point p, the Riemannian distance d_R satisfies:

$$d_R(q, q') = \|x(q) - x(q')\|_{g_p} + o(\|x(q)\|_{g_p} + \|x(q')\|_{g_p}),$$

where $\| \cdot \|_{g_p}$ is the Euclidean norm induced by the value g_p of the metric g at p. This formula has two consequences: first, it shows that the Riemannian distance behaves at the first-order as the Euclidean distance associated with $\| \cdot \|_{g_p}$; secondly, the norm $\| \cdot \|_{g_p}$ gives explicit estimates of d_R near p, such as

$$\frac{1}{C} \|x(q)\|_{g_p} \leq d_R(p, q) \leq C \|x(q)\|_{g_p}.$$

In sub-Riemannian geometry, the two properties above hold, but do not depend on the same function: the first-order behaviour near p is characterized by the distance $\widehat{d}_p$ defined by a nilpotent approximation at p, whereas explicit local estimates of $d(p, \cdot)$ are given by the pseudo-norm $\| \cdot \|_p$ at p defined in (2.5). We first present the latter estimates, often referred to as the "Ball-Box Theorem", and then the first-order expansion of d in Theorem 2.2.

Theorem 2.1 *The following statement holds if and only if $z = (z_1, \ldots, z_n)$ is a system of privileged coordinates at p: there exist constants C_p and $\varepsilon_p > 0$ such that, if $d(p, q) < \varepsilon_p$, then*

$$\frac{1}{C_p} \|z(q)\|_p \leq d(p, q) \leq C_p \|z(q)\|_p, \tag{2.8}$$

where $z(q)$ denotes the coordinates of q and $\| \cdot \|_p$ the pseudo-norm at p.

Corollary 2.1 (Ball-Box Theorem) *Expressed in a given system of privileged coordinates, the sub-Riemannian balls $B(p, \varepsilon)$ satisfy, for $\varepsilon < \varepsilon_p$,*

$$\mathrm{Box}\Big(\frac{1}{C_p}\varepsilon\Big) \subset B(p, \varepsilon) \subset \mathrm{Box}\big(C_p\varepsilon\big),$$

where $\mathrm{Box}(\varepsilon) = [-\varepsilon^{w_1}, \varepsilon^{w_1}] \times \cdots \times [-\varepsilon^{w_n}, \varepsilon^{w_n}].$

The Ball-Box Theorem is stated in different papers, often under the hypothesis that the point p is regular. To our knowledge, two valid proofs exist, the ones in [22] and in [4]. The result also appears without proof in [15] and in [9], and with erroneous proofs in [19] and in [21].

We present here a proof adapted from the one of Bellaïche [(our is much simpler because Bellaïche actually proves a more general result, namely (2.16)]. Basically, the idea is to compare the distances d and $\widehat{d}$. The main step is Lemma 2.2 below, which is essential in other respects to explain the role of nilpotent approximations in control theory.

Fix a point $p \in M$, and a system of privileged coordinates $z = (z_1, \ldots, z_n)$ at p. Through these coordinates we identify a neighbourhood of p in M with a neighbourhood of 0 in $\mathbb{R}^n$. As in the previous section we denote by $\widehat{X}_1, \ldots, \widehat{X}_m$ the homogeneous nilpotent approximation of $X_1, \ldots, X_m$ at p (associated with the given system of privileged coordinates) and by $\widehat{d}$ the induced sub-Riemannian distance on $\mathbb{R}^n$. Recall also that $r = w_n$ denotes the degree of nonholonomy at p.

Lemma 2.2 *There exist constants C and $\varepsilon > 0$ such that, for any $z^0 \in \mathbb{R}^n$ and any $t \in \mathbb{R}^+$ satisfying $\tau = \max(\|z^0\|_p, t) < \varepsilon$, we have*

$$\|z(t) - \widehat{z}(t)\|_p \leq C\tau t^{1/r},$$

where $z(\cdot)$ and $\widehat{z}(\cdot)$ are trajectories of the nonholonomic systems defined respectively by $X_1, \ldots, X_m$ and $\widehat{X}_1, \ldots, \widehat{X}_m$, starting at the same point z^0, associated with the same control function $u(\cdot)$, and satisfying $\|u(t)\| = 1$ a.e.

Proof The first step is to prove that $\|z(t)\|_p$ and $\|\widehat{z}(t)\|_p \leq Cst\ \tau$ for small enough τ, where Cst is a constant. Let us do it for $z(t)$, the proof being exactly the same for $\widehat{z}(t)$.

The equation of the control system associated with $X_1, \ldots, X_m$ is

$$\dot{z}_j = \sum_{i=1}^{m} u_i\big(f_{ij}(z) + r_{ij}(z)\big), \quad j = 1, \ldots, n,$$

where $f_{ij}(z) + r_{ij}(z)$ is of order $\leq w_j - 1$ at 0. Thus, for $j = 1, \ldots, n$ and $i = 1, \ldots, m$, $|f_{ij}(z) + r_{ij}(z)| \leq Cst\ \|z\|_p^{w_j-1}$ when $\|z\|_p$ is small enough. Note that, along the trajectory $z(t)$, $\|z\|_p$ is small when τ is. Since $\|u(t)\| = 1$ a.e., we get:

$$|\dot{z}_j| \le Cst \, \|z\|_p^{w_j-1}. \tag{2.9}$$

To integrate this inequality, choose an integer N such that all N/w_j are even integers and set $\|z\|_N = \left(\sum_{i=1}^n |z_i|^{N/w_i}\right)^{1/N}$. The function $\|z\|_N$ is equivalent to $\|z\|_p$ in the norm sense, and it is differentiable except at the origin. Inequality (2.9) implies $\frac{d}{dt}\|z\|_N \le Cst$, and then, by integration,

$$\|z(t)\|_N \le Cst \, t + \|z(0)\|_N \le Cst \, \tau.$$

The functions $\|z\|_N$ and $\|z\|_p$ being equivalent, we obtain, for a trajectory starting at z^0, $\|z(t)\|_p \le Cst \, \tau$ when τ is small enough.

The second step is to prove $|z_j(t) - \widehat{z}_j(t)| \le Cst \, \tau^{w_j} t$. The function $z_j - \widehat{z}_j$ satisfies the differential equation

$$\dot{z}_j - \dot{\widehat{z}}_j = \sum_{i=1}^m u_i \left(f_{ij}(z) - f_{ij}(\widehat{z}) + r_{ij}(z) \right),$$

$$= \sum_{i=1}^m u_i \left(\sum_{\{k\,:\,w_k<w_j\}} (z_k - \widehat{z}_k) Q_{ijk}(z,\widehat{z}) + r_{ij}(z) \right),$$

where $Q_{ijk}(z,\widehat{z})$ is a homogeneous polynomial of weighted degree $w_j - w_k - 1$. For $\|z\|_p$ and $\|\widehat{z}\|_p$ small enough, we have

$$|r_{ij}(z)| \le Cst \, \|z\|_p^{w_j} \quad \text{and} \quad |Q_{ijk}(z,\widehat{z})| \le Cst \, (\|z\|_p + \|\widehat{z}\|_p)^{w_j-w_k-1}.$$

Using the inequalities of the first step, we obtain finally, for τ small enough,

$$|\dot{z}_j(t) - \dot{\widehat{z}}_j(t)| \le Cst \sum_{\{k\,:\,w_k<w_j\}} |z_k(t) - \widehat{z}_k(t)| \tau^{w_j-w_k-1} + Cst \, \tau^{w_j}. \tag{2.10}$$

This system of inequalities has a triangular form, hence it can be integrated iteratively. For $w_j = 1$, the inequality is $|\dot{z}_j(t) - \dot{\widehat{z}}_j(t)| \le Cst \, \tau$, and so $|z_j(t) - \widehat{z}_j(t)| \le Cst \, \tau t$. By induction, let $j > n_1$ and assume $|z_k(t) - \widehat{z}_k(t)| \le Cst \, \tau^{w_k} t$ for $k < j$. Inequality (2.10) implies

$$|\dot{z}_j(t) - \dot{\widehat{z}}_j(t)| \le Cst \, \tau^{w_j-1} t + Cst \, \tau^{w_j} \le Cst \, \tau^{w_j},$$

and so $|z_j(t) - \widehat{z}_j(t)| \le Cst \, \tau^{w_j} t$.

Finally,

$$\|z(t) - \widehat{z}(t)\|_p \le Cst \, \tau (t^{1/w_1} + \cdots + t^{1/w_n}) \le Cst \, \tau t^{1/r},$$

which completes the proof of the lemma. $\qquad\square$

Proof [of Theorem 2.1] Observe first that, by definition of the order, a system of coordinates z is privileged if and only if $d(p, q) \geq Cst \, \|z(q)\|_p$. What remains to prove is that, if z are privileged coordinates, then $d(p, q) \leq Cst \, \|z(q)\|_p$.

As above, we identify a neighbourhood of p in M with a neighbourhood of 0 in $\mathbb{R}^n$ through the coordinates z. We will show that, for $\|z^0\|_p$ small enough,

$$d(0, z^0) \leq 2\widehat{d}(0, z^0),$$

and so $d(0, z^0) \leq Cst \, \|z^0\|_p$ by Lemma 2.1. This will prove Theorem 2.1.

Fix $z^0 \in \mathbb{R}^n$, $\|z^0\|_p < \varepsilon$. Let $\widehat{z}_0(t)$, $t \in [0, T_0]$, be a minimizing curve for $\widehat{d}$, having velocity one, and joining z^0 to 0. According to Remark 1.2, such a curve exists, and there exists a control $u_0(\cdot)$ associated with $\widehat{z}_0$ such that $\|u_0(t)\| = 1$ a.e. Moreover, $T_0 = \widehat{d}(0, z^0)$.

Let $z_0(t)$, $t \in [0, T_0]$, be the trajectory of the control system associated with $X_1, \ldots, X_m$ starting at z^0 and defined by $u_0(\cdot)$. We have $\mathrm{length}(z_0(\cdot)) \leq T_0$. Set $z^1 = z_0(T_0)$. By Lemma 2.2,

$$\|z^1\|_p = \|z_0(T_0) - \widehat{z}_0(T_0)\|_p \leq C\tau T_0^{1/r},$$

where $\tau = \max(\|z^0\|_p, T_0)$. By Lemma 2.1, $T_0 = \widehat{d}(0, z^0)$ satisfies $T_0 \geq \|z^0\|_p/C'$, so $\tau \leq C'T_0$, and

$$\widehat{d}(0, z^1) \leq C'\|z^1\|_p \leq C''\widehat{d}(0, z^0)^{1+1/r},$$

with $C'' = C'^2 C$.

Choose now $\widehat{z}_1(t)$, $t \in [0, T_1]$, a minimizing curve for $\widehat{d}$ of velocity one joining z^1 to 0. There exists a control $u_1(\cdot)$ associated with $\widehat{z}_1$ such that $\|u_1(t)\| = 1$ a.e. Let $z_1(t)$, $t \in [0, T_1]$, be the trajectory of the control system associated with $X_1, \ldots, X_m$ starting at z^1 and defined by $u_1(\cdot)$. Set $z^2 = z_1(T_1)$. As previously, we have $\mathrm{length}(z_1(\cdot)) = \widehat{d}(0, z^1)$ and $\widehat{d}(0, z^2) \leq C''\widehat{d}(0, z^1)^{1+1/r}$.

Repeating this construction, we obtain a sequence of points $z^0, z^1, z^2, \ldots$ such that $\widehat{d}(0, z^{k+1}) \leq C''\widehat{d}(0, z^k)^{1+1/r}$, and a sequence of trajectories $z_k(\cdot)$ joining z^k to z^{k+1} of length equal to $\widehat{d}(0, z^k)$.

Taking $\|z^0\|_p$ small enough, we can assume $C''\widehat{d}(0, z^0)^{1/r} \leq 1/2$. We have then $\widehat{d}(0, z^1) \leq \widehat{d}(0, z^0)/2, \ldots, \widehat{d}(0, z^k) \leq \widehat{d}(0, z^0)/2^k, \ldots$ Consequently, z^k tends to 0 as $k \to +\infty$. Putting end to end the curves $z_k(\cdot)$ gives a trajectory joining z^0 to 0 of length $\widehat{d}(0, z^0) + \widehat{d}(0, z^1) + \cdots \leq 2\widehat{d}(0, z^0)$. This implies $d(0, z^0) \leq 2\widehat{d}(0, z^0)$, and the proof is complete. $\qquad\square$

Now, the distance $\widehat{d}$ on $\mathbb{R}^n$ induces a distance $\widehat{d}_p$ on a neighbourhood of p in M by setting $\widehat{d}_p(q, q') = \widehat{d}(z(q), z(q'))$. This distance gives the first-order term in the expansion of $d(p, \cdot)$.

Theorem 2.2 *On a neighbourhood of p in M there holds*

$$d(p, q) = \widehat{d}_p(p, q) \left(1 + O\left(\widehat{d}_p(p, q)\right)\right).$$

Remark 2.3 By Theorem 2.1 and Lemma 2.1, when $d(p, q)$ is small enough we get the estimate

$$\frac{1}{C}\widehat{d}_p(p, q) \le d(p, q) \le C\widehat{d}_p(p, q), \tag{2.11}$$

where C is some positive constant. Theorem 2.2 essentially states that this constant can be chosen arbitrarily close to 1.

Proof Fix $\delta > 0$. We have to prove that there exists $\varepsilon > 0$ such that, if $d(p, q) < \varepsilon$, then

$$(1 - \delta)\widehat{d}_p(p, q) \le d(p, q) \le (1 + \delta)\widehat{d}_p(p, q). \tag{2.12}$$

Let q be a point in M. Setting $z^0 = z(q)$ in the proof of Theorem 2.1 furnishes a trajectory joining q to 0 whose length is equal to $\sum_{k=0}^{\infty} \widehat{d}(0, z^k)$, the points z^k being such that $\widehat{d}(0, z^{k+1}) \le C''\widehat{d}(0, z^k)^{1+1/r}$.

From (2.11), there exists $\varepsilon > 0$ such that $d(p, q) < \varepsilon$ implies $C''\widehat{d}(0, z^0)^{1/r} \le \delta/(1 + \delta)$. In this case the length of the trajectory from q to 0 is not greater than $(1 + \delta)\widehat{d}(0, z^0)$ and we have

$$d(p, q) \le (1 + \delta)\widehat{d}_p(p, q),$$

since $\widehat{d}(0, z^0) = \widehat{d}_p(p, q)$.

To prove the other inequality in (2.12), we use the same argument but reverse the role of d and $\widehat{d}$. Again, we identify a neighbourhood of p in M with a neighbourhood of 0 in $\mathbb{R}^n$ through the coordinates z. Let $z_0(t), t \in [0, T_0]$, be a minimizing curve for d of velocity one joining z^0 to 0, and let $u_0(\cdot)$ be a control associated with $z_0(\cdot)$ such that $\|u_0(t)\| = 1$ a.e. We have $T_0 = d(0, z^0)$. Let $\widehat{z}_0(t), t \in [0, T_0]$, be the trajectory of the control system associated with $\widehat{X}_1, \ldots, \widehat{X}_m$ starting at z^0 and defined by the control $u_0(\cdot)$. In particular, $\text{length}\left(\widehat{z}_0(\cdot)\right) \le T_0$.

Set $z^1 = z_0(T_0)$. By Lemma 2.2,

$$\|z^1\|_p = \|z_0(T_0) - \widehat{z}_0(T_0)\|_p \le C\tau T_0^{1/r},$$

where $\tau = \max(\|z^0\|_p, T_0)$. Theorem 2.1 implies $\tau \le C_p T_0$, and

$$d(0, z^1) \le C_p\|z^1\|_p \le C''d(0, z^0)^{1+1/r},$$

with $C'' = C_p^2 C$.

Repeating this construction gives a trajectory of $\widehat{X}_1, \ldots, \widehat{X}_m$ joining q to p whose length is equal to $\sum_{k=0}^{\infty} d(0, z^k)$, where $d(0, z^{k+1}) \le C''d(0, z^k)^{1+1/r}$.

For $d(p, q)$ small enough, we have $C''d(0, z^0)^{1/r} \leq \delta/(1 - \delta)$ and the trajectory from q to 0 is of length $\leq 1/(1 - \delta)d(0, z^0)$, which leads to

$$\widehat{d}_p(p, q) \leq \frac{1}{(1 - \delta)}d(p, q).$$

This completes the proof. $\square$

2.2.2 Uniform Estimates

In the preceding sections we have built notions of privileged coordinates and nilpotent approximations at a given point p, as well as distance estimates from this point. We now study in which extent these quantities vary continuously as the point p varies.

Let first define what we mean by continuity of a system of privileged coordinates.

Definition 2.8 A *continuously varying system of privileged coordinates* on an open subset Ω of M is a mapping Φ taking values in $\mathbb{R}^n$, defined and continuous on a neighbourhood of the set $\{(q, q), q \in \Omega\} \subset \Omega \times \Omega$, and so that the partial mapping $\Phi(p, \cdot)$ is a system of privileged coordinates at p.

In a neighbourhood of a regular point $\bar{p}$, such a continuously varying system of privileged coordinates always exist. Indeed, all the examples of construction of privileged coordinates given in Sect. 2.1.2 are based on adapted frames. And, the point $\bar{p}$ being regular, an adapted frame $Y_1, \ldots, Y_n$ at $\bar{p}$ will be an adapted frame at every p in a neighbourhood Ω of $\bar{p}$. It is then easy to see that using the same adapted frame at every $p \in \Omega$ will give privileged coordinates $\Phi(p, \cdot)$ varying continuously with p. For instance, if $\Phi(p, q) = z$ is defined by

$$q = \exp(z_1 Y_1 + \cdots + z_n Y_n)(p),$$

the continuity of Φ results from the Implicit Function Theorem.

On the other hand, the existence of a continuous varying system of privileged coordinates is not ensured near a singular point. It is not excluded either, as shown by the following example.

Example 2.9 (Martinet case) Consider the Martinet case, described in Example 2.5. There exists singular points, those are the points of the plane $\{x = 0\}$. It is however easy to check that the following map Φ defines a continuous varying system of privileged coordinates on $\mathbb{R}^3$,

$$\Phi\left((\bar{x}, \bar{y}, \bar{z}), (x, y, z)\right) = \left(x - \bar{x}, y - \bar{y}, z - \bar{z} - \frac{\bar{x}^2}{2}(y - \bar{y})\right),$$

the third coordinate being of order 2 if $\bar{x} \neq 0$, and of order 3 if $\bar{x} = 0$.

Definition 2.9 A *nilpotent approximation of* $(X_1, \ldots, X_m)$ *on an open subset* Ω *of* M is a mapping $\mathscr{A}$ which associates, with every $p \in \Omega$, a nilpotent approximation of $(X_1, \ldots, X_m)$ at p,

$$\mathscr{A}(p) = (\widehat{X}_1^p, \ldots, \widehat{X}_m^p).$$

We say that $\mathscr{A}$ is *continuous* if the mapping $(p, q) \mapsto \mathscr{A}(p)(q)$ is well-defined and continuous on a neighbourhood of the set $\{(p, p), p \in \Omega\} \subset \Omega \times \Omega$.

Recall (Definition 2.7) that a nilpotent approximation $(\widehat{X}_1^p, \ldots, \widehat{X}_m^p)$ at p is the homogeneous nilpotent approximation at p associated with some privileged coordinates z at p, that is, $\widehat{X}_i^p$ is the homogeneous part of weighted degree -1 in the Taylor expansion of X_i expressed in coordinates z. Now, if the weights are constant on Ω and if there exists a continuous varying system of privileged coordinates on Ω, the resulting nilpotent approximation on Ω is continuous. For every regular point $p \in M$ both conditions are satisfied near p, which ensures the existence of a continuous nilpotent approximation in a neighbourhood of p.

Near a singular point on the contrary, the weights are not constants and a priori there do not exist continuous nilpotent approximations.

Assume now that a continuous varying system of privileged coordinates and a continuous nilpotent approximation are given on an open set $\Omega \subset M$ containing only regular points. It is just a matter of routine to check in the computations of Sect. 2.2.1 that the constants C_p, ε_p of Theorem 2.1 and C, ε of Lemma 2.2 may be chosen as continuous functions of p. More precisely, we have the following result, which summarizes the discussion above.

Theorem 2.3 *Let* $\bar{p} \in M$ *be a regular point. There exists an open neighbourhood* Ω *of* $\bar{p}$, *a continuous varying system of privileged coordinates* Φ *on* Ω, *a continuous nilpotent approximation* $\mathscr{A}$ *on* Ω, *and two positive continuous functions* $\varepsilon(\cdot), C(\cdot)$, *such that the following properties are satisfied.*

For every pair $(p, q) \in \Omega \times \Omega$ *with* $d(p, q) < \varepsilon(p)$, *there holds,*

$$\frac{1}{C(p)} \|z(q)\|_p \leq d(p, q) \leq C(p)\|z(q)\|_p, \tag{2.13}$$

where $z = \Phi(p, \cdot)$. *Moreover, for every control* $u(\cdot)$ *satisfying* $\|u\|_{L^1} < \varepsilon(p)$, *we have*

$$\|z(\gamma(T; q, u)) - z(\hat{\gamma}(T; q, u))\|_p \leq C(p) \max \left(\|z(q)\|_p, \|u\|_{L^1} \right) \|u\|_{L^1}^{1/r}, \tag{2.14}$$

where r *is the degree of nonholonomy at* p, *and* $\hat{\gamma}(\cdot; q, u)$ *is a trajectory of the nonholonomic system defined by* $\mathscr{A}(p) = (\widehat{X}_1^p, \ldots, \widehat{X}_m^p)$.

This result underlines the importance of the case where there are no singular points.

Definition 2.10 A manifold M equipped with vector fields $X_1, \ldots, X_m$ is said to be *equiregular* if every point in M is regular w.r.t. $X_1, \ldots, X_m$.

As a direct consequence of Theorem 2.3 we have an estimate of the volume of small sub-Riemannian balls in equiregular manifolds. Assume that M is an oriented manifold with a volume form ω and let vol_ω be the measure on M associated with ω, i.e. $\mathrm{vol}_\omega(A) = \int_A \omega$ for any measurable subset $A \subset M$.

Corollary 2.2 *Assume that M is equiregular. Given a compact set $K \subset M$ there exist positive constants C and ε_0 such that, for all $q \in K$ and $\varepsilon < \varepsilon_0$,*

$$\frac{1}{C}\varepsilon^Q \leq \mathrm{vol}_\omega(B(q, \varepsilon)) \leq C\varepsilon^Q, \tag{2.15}$$

where $Q = \sum_{i=1}^n w_i(q)$ does not depend on q.

Uniform Ball-Box Theorem

Near a singular point, it is no longer possible to have continuous functions $\varepsilon(\cdot)$, $C(\cdot)$ like the ones of Theorem 2.3. In particular, if (p_n) is a sequence of regular points converging to a singular point p (this is possible since regular points are dense in M), the sequence $\varepsilon(p_n)$ tends to zero whereas the constant ε_p of Theorem 2.1 is not equal to zero. It is however possible to obtain some uniform estimates through a process of desingularization. We just mention here the results, we refer the reader to [16] for the proofs.

Let $K \subset M$ be a compact set and $r_{\max}$ be the maximum of the degree of non-holonomy on K (as noticed in Sect. 2.1.2 $r_{\max}$ is finite). We assume that M is an oriented manifold with a volume form ω.

Let $\mathfrak{X}$ be the set of n-tuples $\mathbf{X} = (X_{I_1}, \ldots, X_{I_n})$ of brackets of length $|I_i| \leq r_{\max}$. It is a finite subset of $Lie(X_1, \ldots, X_m)^n$. Given $q \in K$ and $\varepsilon > 0$ we define a function $f_{q,\varepsilon} : \mathfrak{X} \to \mathbb{R}$ by

$$f_{q,\varepsilon}(\mathbf{X}) = \left| \omega_q \left(X_{I_1}(q)\varepsilon^{|I_1|}, \ldots, X_{I_n}(q)\varepsilon^{|I_n|} \right) \right|.$$

We say that $\mathbf{X} \in \mathfrak{X}$ is an *adapted frame* at (q, ε) if it achieves the maximum of $f_{q,\varepsilon}$ on $\mathfrak{X}$.

The values at q of an adapted frame at (q, ε) clearly form a basis of $T_q M$. Moreover, q being fixed, the adapted frames at (q, ε) are adapted frames at q for ε small enough.

Theorem 2.4 (Uniform Ball-Box theorem [16]) *There exist positive constants C and ε_0 such that, for $q \in K$, $\varepsilon < \varepsilon_0$, and any adapted frame $\mathbf{X}$ at (q, ε), there holds*

$$\mathrm{Box}_{\mathbf{X}}(q, \tfrac{1}{C}\varepsilon) \subset B(q, \varepsilon) \subset \mathrm{Box}_{\mathbf{X}}(q, C\varepsilon),$$

where $\mathrm{Box}_{\mathbf{X}}(q, \varepsilon) = \{\exp(x_1 X_{I_1}) \circ \cdots \circ \exp(x_n X_{I_n})(q) : |x_i| \leq \varepsilon^{|I_i|}, 1 \leq i \leq n\}$.

Of course, when the point q is fixed, the estimate above is equivalent to the one of the Ball-Box theorem for ε smaller than some $\varepsilon_1(q) > 0$. However, the main difference is that here ε_0 does not depend on q, whereas in the Ball-Box theorem $\varepsilon_1 = \varepsilon_1(q)$ can be infinitely close to 0 as q varies.

As a consequence of the Uniform Ball-Box theorem, we obtain an estimate of the volume of small sub-Riemannian balls that generalizes Corollary 2.2.

Corollary 2.3 *There exist positive constants C and ε_0 such that, for all $q \in K$ and $\varepsilon < \varepsilon_0$,*

$$\frac{1}{C} \max_{\mathbf{X}} f_{q,\varepsilon}(\mathbf{X}) \leq \mathrm{vol}_\omega(B(q,\varepsilon)) \leq C \max_{\mathbf{X}} f_{q,\varepsilon}(\mathbf{X}),$$

the maximum of $f_{q,\varepsilon}(\mathbf{X}) = \left| \omega_q\left(X_{I_1}(q)\varepsilon^{|I_1|}, \ldots, X_{I_n}(q)\varepsilon^{|I_n|}\right) \right|$ being taken over all n-tuples $\mathbf{X} = (X_{I_1}, \ldots, X_{I_n})$ of brackets of length $|I_i| \leq r_{\max}$.

2.3 Application to Carnot-Carathéodory Spaces

The manifold M endowed with the sub-Riemannian distance d defines a metric space (M, d) which is called a *Carnot-Carathéodory space*. The first-order theory introduced previously has a metric interpretation and will allow us to describe the local structure of a Carnot-Carathéodory space.

2.3.1 Tangent Structure to Carnot-Carathéodory Spaces

In describing the tangent space to a manifold, we essentially look at smaller and smaller neighbourhoods of a given point, the manifold being fixed. Equivalently, we can look at a fixed neighbourhood, but expanding the manifold. As noticed by Gromov, this idea can be used to define a notion of tangent space for a general metric space.

If X is a metric space with distance d, we define $\lambda\mathsf{X}$, for $\lambda > 0$, to be the metric space with same underlying set as X and distance λd. A *pointed metric space* (X, x) is a metric space with a distinguished point x.

Loosely speaking, a metric tangent space to the metric space X at x is a pointed metric space $(C_x\mathsf{X}, y)$ such that

$$(C_x\mathsf{X}, y) = \lim_{\lambda \to +\infty} (\lambda\mathsf{X}, x).$$

Of course, for this definition to make sense, we have to define the limit of pointed metric spaces.

Let us first define the Gromov-Hausdorff distance between metric spaces. Recall that, in a metric space X, the Hausdorff distance $\mathrm{H} - \mathrm{dist}(A, B)$ between two subsets

A and B of X is the infimum of $\rho > 0$ such that any point of A is within a distance ρ of B and any point of B is within a distance ρ of A. The Gromov-Hausdorff distance $\mathrm{GH} - \mathrm{dist}(X, Y)$ between two metric spaces X and Y is the infimum of Hausdorff distances $\mathrm{H} - \mathrm{dist}(i(\mathsf{X}), j(\mathsf{Y}))$ over all metric spaces Z and all isometric embeddings $i : \mathsf{X} \to \mathsf{Z}, j : \mathsf{Y} \to \mathsf{Z}$.

Thanks to Gromov-Hausdorff distance, one can define the notion of limit of a sequence of pointed metric spaces: (X_n, x_n) converge to (X, x) if, for any positive r,

$$\mathrm{GH} - \mathrm{dist}\big(B^{\mathsf{X}_n}(x_n, r), B^{\mathsf{X}}(x, r)\big) \to 0 \quad \text{as } n \to +\infty$$

where $B^{\mathsf{Y}}(y, r)$ endowed with the distance of Y is considered as a metric space. Note that all pointed metric spaces isometric to (X, x) are also limit of (X_n, x_n). However the limit is unique up to an isometry provided the closed balls around the distinguished point are compact [3, Sect. 7.4].

Finally, one says that $(\mathsf{X}_\lambda, x_\lambda)$ converge to (X, x) when $\lambda \to \infty$ if, for every sequence λ_n, $(\mathsf{X}_{\lambda_n}, x_{\lambda_n})$ converge to (X, x).

Definition 2.11 A pointed metric space $(C_x\mathsf{X}, y)$ is a *metric tangent space* to the metric space X at x if $(\lambda\mathsf{X}, x)$ converge to $(C_x\mathsf{X}, y)$ as $\lambda \to +\infty$. If it exists, it is unique up to an isometry provided the closed balls around x in $(\lambda\mathsf{X}, x)$ are compact.

For a Riemannian metric space (M, d_R) induced by a Riemannian metric g on a manifold M, metric tangent spaces at a point p exist and are isometric to the Euclidean space (T_pM, g_p), that is, the standard tangent space endowed with the scalar product defined by the quadratic form g_p.

For a Carnot-Carathéodory space (M, d), the metric tangent space is given by the nilpotent approximation.

Theorem 2.5 *A Carnot-Carathéodory space (M, d) admits metric tangent spaces (C_pM, y) at every point $p \in M$. The space C_pM is itself a Carnot-Carathéodory space isometric to $(\mathbb{R}^n, \widehat{d})$, where $\widehat{d}$ is the sub-Riemannian distance associated with a homogeneous nilpotent approximation at p.*

This theorem, due to Bellaïche, is a consequence of a strong version of Theorem 2.1: for q and q' in a neighbourhood of p,

$$|d(q, q') - \widehat{d}(q, q')| \leq Cst\,\widehat{d}(p, q)d(q, q')^{1/r}. \tag{2.16}$$

In these notes, we present neither the proof of this result, nor the one of Theorem 2.5, and we refer the reader to [4, 21].

Remark 2.4 Recall that $\widehat{d}$ is not intrinsic to the frame $(X_1, \ldots, X_m)$. Thus Theorem 2.5 does not provide an intrinsic characterization of the metric tangent space. Such characterizations exist for sub-Riemannian manifolds (M, D, g_R) in [7, 20], and the latter could easily be adapted to the case of a sub-Riemannian geometry associated with a nonholonomic system. However these constructions are intrinsic to the differentiable manifold M equipped with the sub-Riemannian structure (D, g_R), or to

M equipped with the frame $(X_1, \ldots, X_m)$, not to the metric space (M, d). To our knowledge, the problem of finding a characterization of the metric tangent space $C_p M$ depending only on the Carnot-Carathéodory space (M, d) is still open.

What is the algebraic structure of $C_p M$? Of course $C_p M$ is not a linear space in general: for instance, $\widehat{d}$ is homogeneous of degree 1, but with respect to dilations δ_t and not with respect to the usual Euclidean dilations. It appears that $C_p M$ has actually a natural structure of group, or at least of quotient of groups.

Let us introduce the group G_p generated by the diffeomorphisms $\exp(t\widehat{X}_i)$ acting on the left on $\mathbb{R}^n$ (note that every $\exp(t\widehat{X}_i)$ is a global diffeomorphism on $\mathbb{R}^n$ since $\widehat{X}_i$ is a complete vector field). Since $\mathfrak{g}_p = Lie(\widehat{X}_1, \ldots, \widehat{X}_m)$ is a nilpotent Lie algebra, G_p is a simply connected Lie group and has $\mathfrak{g}_p$ as its Lie algebra, that is $G_p = \exp(\mathfrak{g}_p)$. This Lie algebra $\mathfrak{g}_p$ splits into homogeneous components

$$\mathfrak{g}_p = \mathfrak{g}^{-1} \oplus \cdots \oplus \mathfrak{g}^{-r},$$

where $\mathfrak{g}^{-s}$ is the set of homogeneous vector fields of degree $-s$, and so $\mathfrak{g}_p$ is a graded Lie algebra. The first component $\mathfrak{g}^{-1} = \mathrm{span}\langle \widehat{X}_1, \ldots, \widehat{X}_m \rangle$ generates $\mathfrak{g}_p$ as a Lie algebra. All these properties imply that G_p is what we call a Carnot group.

Definition 2.12 A *Carnot group* is a simply connected Lie group, such that the associated Lie algebra is graded, nilpotent, and generated by its first component.

Note that the dilations δ_t act on $\mathfrak{g}_p$ as a multiplication by t^{-s} on $\mathfrak{g}^{-s}$. This action extends to G_p by the exponential mapping.

Example 2.10 (Heisenberg group) The simplest non Abelian Carnot group is the *Heisenberg group* $\mathbb{H}^3$ which is the connected and simply connected Lie group whose Lie algebra satisfies

$$\mathfrak{g} = \mathfrak{g}^{-1} \oplus \mathfrak{g}^{-2}, \qquad \text{with } \dim \mathfrak{g}^{-1} = 2.$$

As a consequence, $\dim \mathbb{H}^3 = 3$. Choosing a basis $X, Y, Z = [X, Y]$ of $\mathfrak{g}$, we define coordinates on $\mathbb{H}^3$ by the exponential mapping

$$(x, y, z) \mapsto \exp(xX + yY + zZ).$$

By the Campbell-Hausdorff formula (see Sect. A.1), the law group on $\mathbb{H}^3$ in these coordinates is

$$(x, y, z) \cdot (x', y', z') = (x + x', y + y', z + z' + \frac{1}{2}(xy' - x'y)),$$

which is homogeneous with respect to the dilation $\delta_t(x, y, z) = (tx, ty, t^2 z)$.

Finally, denote by X_1, X_2 the left-invariant vector fields on $\mathbb{H}^3$ whose values at the identity are respectively X and Y. In coordinates (x, y, z), these vector fields write as

$$X_1 = \partial_x - \frac{y}{2}\partial_z \quad \text{and} \quad X_2 = \partial_y + \frac{x}{2}\partial_z,$$

which are the vector fields of what we have called the Heisenberg case in Examples 2.3 and 2.4.

Let $\widehat{\xi}_1, \ldots, \widehat{\xi}_m$ be the right-invariant vector fields on G_p such that $\widehat{\xi}_i(\mathrm{id}) = \widehat{X}_i$, where id is the identity of G_p. Equivalently,

$$\widehat{\xi}_i(g) = \frac{d}{dt}\big[\exp(t\widehat{X}_i)g\big]\big|_{t=0}.$$

With $(\widehat{\xi}_1, \ldots, \widehat{\xi}_m)$ is associated a right-invariant sub-Riemannian metric and a sub-Riemannian distance d_{G_p} on G_p.

The action of G_p on $\mathbb{R}^n$ is smooth and transitive. Indeed, for every $x \in \mathbb{R}^n$, the orbit of x under the action of G_p is the set

$$\left\{\exp(t_{i_1}\widehat{X}_{i_1}) \circ \cdots \circ \exp(t_{i_k}\widehat{X}_{i_k})(x) \ : \ k \in \mathbb{N}, \ t_{i_j} \in \mathbb{R}, \ i_j \in \{1, \ldots, m\}\right\}.$$

By Chow-Rashevsky's theorem (or more precisely by Remark 1.6), this set is the whole $\mathbb{R}^n$ since $(\widehat{X}_1, \ldots, \widehat{X}_m)$ satisfies Chow's Condition on $\mathbb{R}^n$ (Lemma 2.1).

To understand the algebraic structure of C_pM we will use the following standard result on transitive action of Lie groups (see for instance [18, Theorem 9.24]).

Theorem 2.6 *Let G be a Lie group acting on the left smoothly and transitively on a manifold M. Let $q \in M$ and H be the isotropy subgroup of q which is defined by $H = \{g \in G \ : \ g \cdot q = q\}$. Then H is a closed subgroup of G, the left coset space G/H is a manifold of dimension $\dim G - \dim H$, and the map $F : G/H \to M$ defined by $F(gH) = g \cdot q$ is an equivariant diffeomorphism.*

Let H_p be the isotropy subgroup of $0 \in \mathbb{R}^n$ under the action of G_p. According to Theorem 2.6, the map $\phi_p : G_p \to \mathbb{R}^n$, $\phi_p(g) = g(0)$, induces a diffeomorphism

$$\psi_p : G_p/H_p \to \mathbb{R}^n, \qquad \psi_p(gH_p) = g(0).$$

Observe that H_p is invariant under dilations, since $\delta_t g(\delta_t x) = \delta_t(g(x))$. Hence H_p is connected and simply connected, and so $H_p = \exp(\mathfrak{h}_p)$, where $\mathfrak{h}_p$ is the Lie sub-algebra of $\mathfrak{g}_p$ containing the vector fields vanishing at 0,

$$\mathfrak{h}_p = \{Z \in \mathfrak{g}_p \ : \ Z(0) = 0\}.$$

As $\mathfrak{g}_p$, $\mathfrak{h}_p$ is invariant under dilations and splits into homogeneous components.

Now, the elements $\widehat{X}_1, \ldots, \widehat{X}_m$ of $\mathfrak{g}_p$ act on the left on G_p/H_p with the notation $\overline{\xi}_1, \ldots, \overline{\xi}_m$,

$$\overline{\xi}_i(gH_p) = \frac{d}{dt}\big[\exp(t\widehat{X}_i)gH_p\big]\big|_{t=0}.$$

These vector fields define a sub-Riemannian metric and a sub-Riemannian distance $\overline{d}$ on G_p/H_p. We clearly have $\psi_{p*}\overline{\xi}_i = \widehat{X}_i$, so ψ_p maps the sub-Riemannian metric on G_p/H_p associated with $(\overline{\xi}_1, \ldots, \overline{\xi}_m)$ to the one on $\mathbb{R}^n$ associated with $(\widehat{X}_1, \ldots, \widehat{X}_m)$. We summarize this construction by the following result.

Theorem 2.7 *The metric tangent space C_pM and $(\mathbb{R}^n, \widehat{d})$ are isometric to the coset space G_p/H_p endowed with the sub-Riemannian distance $\overline{d}$.*

Example 2.11 (Grušin plane) Consider the vector fields $X_1 = \partial_x$ and $X_2 = x\partial_y$ on $\mathbb{R}^2$. The Carnot-Carathéodory space defined by these vector fields is called the *Grušin plane.*

The only nonzero bracket is $X_{[1,2]} = [X_1, X_2] = \partial_y$. Thus, at $p = 0$, the weights are $(1, 2)$, and (x, y) are privileged coordinates. Since X_1 and X_2 are homogeneous with respect to this system of coordinates, we have $\widehat{X}_1 = X_1$ and $\widehat{X}_2 = X_2$. The Lie algebra they generate is

$$\mathfrak{g}_0 = \operatorname{span}(X_1, X_2, X_{[1,2]})$$

which is of dimension 3, and the group $\exp(\mathfrak{g}_0)$ is actually the *Heisenberg group* $\mathbb{H}^3$ (see Example 2.10). The Lie sub-algebra $\mathfrak{h}_0$ of $\mathfrak{g}_0$ containing the vector fields vanishing at 0 is

$$\mathfrak{h}_0 = \operatorname{span}(X_2),$$

which is one-dimensional. Thus the Grušin plane is isometric to $\mathbb{H}^3/\exp(\mathfrak{h}_0)$ endowed with the distance $\overline{d}$.

Example 2.12 (Martinet case) Consider the Martinet case, defined on $\mathbb{R}^3$ by

$$X_1 = \partial_x \quad \text{and} \quad X_2 = \partial_y + \frac{x^2}{2}\partial_z.$$

As noticed in Example 2.5, at $p = 0$ the coordinates (x, y, z) are privileged and by homogeneity $\widehat{X}_1 = X_1$ and $\widehat{X}_2 = X_2$. Moreover the only nonzero bracket are $X_{[1,2]} = x\partial_z$ and $X_{[1,[1,2]]} = \partial_z$. Thus,

$$\mathfrak{g}_0 = \operatorname{span}(X_1, X_2, X_{[1,2]}, X_{[1,[1,2]]})$$

which is of dimension 4. The group $\exp(\mathfrak{g}_0)$ is called the *Engel group*, and is denoted by $\mathbb{E}^4$. The Lie sub-algebra $\mathfrak{h}_0$ of $\mathfrak{g}_0$ containing the vector fields vanishing at 0 is

$$\mathfrak{h}_0 = \operatorname{span}(X_{[1,2]}).$$

When the point p is regular, Theorem 2.7 can be refined thanks to the following result.

Lemma 2.3 *If p is a regular point, then $\dim G_p = n$.*

Proof Let $X_{I_1}, \ldots, X_{I_n}$ be an adapted frame at p. Since p is regular, $X_{I_1}, \ldots, X_{I_n}$ is also an adapted frame near p, and any bracket X_J can be written as

$$X_J(z) = \sum_{\{i\,:\,|I_i| \le |J|\}} a_i(z) X_{I_i}(z),$$

where each a_i is a function of order $\ge |I_i| - |J|$. Taking the homogeneous terms of degree $-|J|$ in this expression, we obtain

$$\widehat{X}_J(z) = \sum_{\{i\,:\,|I_i| = |J|\}} a_i(0) \widehat{X}_{I_i}(z),$$

and so $\widehat{X}_J \in \mathrm{span}\langle \widehat{X}_{I_1}, \ldots, \widehat{X}_{I_n} \rangle$. Thus $\widehat{X}_{I_1}, \ldots, \widehat{X}_{I_n}$ is a basis of $\mathfrak{g}_p$, and so $\dim G_p = n$. $\qquad\square$

As a consequence H_p is of dimension zero in this case. Since H_p is invariant under dilations, $H_p = \{\mathrm{id}\}$, and hence the mapping $\phi_p : G_p \to \mathbb{R}^n$, $\phi_p(g) = g(0)$, is a diffeomorphism. Moreover $\phi_{p*}\widehat{\xi}_i = \widehat{X}_i$, which implies that ϕ_p maps the sub-Riemannian metric on G_p associated with $(\widehat{\xi}_1, \ldots, \widehat{\xi}_m)$ to the one on $\mathbb{R}^n$ associated with $(\widehat{X}_1, \ldots, \widehat{X}_m)$. This gives the following result.[1]

Proposition 2.4 *When p is a regular point, the metric tangent space $C_p M$ and the Carnot-Carathéodory space $(\mathbb{R}^n, \widehat{d})$ are isometric to the Carnot group G_p endowed with the right-invariant sub-Riemannian distance d_{G_p}.*

Thus Carnot groups have the same role in sub-Riemannian geometry as Euclidean spaces have in Riemannian geometry. For this reason they are sometimes referred to as "non Abelian linear spaces": the internal operation—addition—is replaced by the law group and the external operation—multiplication by a real number—by the dilations. Note that, when G_p is Abelian (i.e. commutative) then G_p has a linear structure and the sub-Riemannian metric on G_p is a Euclidean metric.

Example 2.13 (unicycle) In the case of the distance d associated with the unicycle (Examples 1.1 and 2.8), the growth vector is $(2, 3)$ at every point. Hence every point $p \in \mathbb{R}^2 \times \mathscr{S}^1$ is regular, and the Lie algebra generated by the nilpotent approximation satisfies

$$\mathfrak{g}_p = \mathfrak{g}^{-1} \oplus \mathfrak{g}^{-2}, \quad \text{with } \dim \mathfrak{g}^{-1} = 2.$$

As a consequence, $G_p = \mathbb{H}^3$ (see Example 2.10), and so the metric tangent space to $(\mathbb{R}^2 \times \mathscr{S}^1, d)$ at every point p has the structure of the Heisenberg group.

---

[1] This result appeared first in [19], but with an erroneous proof. The presentation given here is inspired from the one of [4].

2.3.2 Hausdorff Dimension

Consider a metric space (M, d) and denote by diam S the diameter of a set $S \subset M$. Let $k \geq 0$ be a real number. For every subset $A \subset M$, we define the *k-dimensional Hausdorff measure* $\mathcal{H}^k$ of A as $\mathcal{H}^k(A) = \lim_{\varepsilon \to 0^+} \mathcal{H}^k_\varepsilon(A)$, where

$$\mathcal{H}^k_\varepsilon(A) = \inf \left\{ \sum_{i=1}^\infty (\text{diam } S_i)^k \ : \ A \subset \bigcup_{i=1}^\infty S_i, \ S_i \text{ closed set, } \text{diam } S_i \leq \varepsilon \right\}.$$

We also define the *k-dimensional spherical Hausdorff measure* $\mathcal{S}^k$ of A as $\mathcal{S}^k(A) = \lim_{\varepsilon \to 0^+} \mathcal{S}^k_\varepsilon(A)$, where

$$\mathcal{S}^k_\varepsilon(A) = \inf \left\{ \sum_{i=1}^\infty (\text{diam } S_i)^k \ : \ A \subset \bigcup_{i=1}^\infty S_i, \ S_i \text{ is a ball, } \text{diam } S_i \leq \varepsilon \right\}.$$

In the Euclidean space $\mathbb{R}^n$, k-dimensional Hausdorff measures are often defined as $2^{-k}\alpha(k)\mathcal{H}^k$ and $2^{-k}\alpha(k)\mathcal{S}^k$, where $\alpha(k)$ is defined from the usual gamma function as $\alpha(k) = \Gamma(\frac{1}{2})^k / \Gamma(\frac{k}{2} + 1)$. This normalization factor is necessary for the n-dimensional Hausdorff measure and the Lebesgue measure to coincide on $\mathbb{R}^n$.

For a given set $A \subset M$, $\mathcal{H}^k(A)$ is a decreasing function of k, infinite when k is smaller than a certain value, and zero when k is greater than this value. We call *Hausdorff dimension* of A the real number

$$\dim_{\mathcal{H}} A = \sup\{k \ : \ \mathcal{H}^k(A) = \infty\} = \inf\{k \ : \ \mathcal{H}^k(A) = 0\}.$$

Note that $\mathcal{H}^k \leq \mathcal{S}^k \leq 2^k \mathcal{H}^k$, so the Hausdorff dimension can be defined equivalently from Hausdorff or spherical Hausdorff measures.

In the case where the metric space (M, d) is a Carnot-Carathéodory space, only few results exist on Hausdorff measures, except for specific cases [1, 12]. The most general result is the following one.

Theorem 2.8 *Let (M, d) be an equiregular Carnot-Carathéodory space and p a point in M. Then the Hausdorff dimension of a small enough ball*
$B(p, r)$ *is* $\dim_{\mathcal{H}} B(p, r) = Q$, *where*

$$Q = \sum_{i=1}^n w_i(p) = \sum_{i \geq 1} i \left(\dim \Delta^i(p) - \dim \Delta^{i-1}(p) \right)$$

does not depend on p. Moreover $\mathcal{H}^Q(B(p, r))$ is finite.

Proof Fix a volume form ω on $B(p, r)$ (it is possible for a small enough r), and denote by vol_ω the associated measure. It results from Corollary 2.2 that, for $q \in B(p, r)$ and ε small enough,

$$\frac{1}{C}\varepsilon^Q \leq \mathrm{vol}_\omega(B(q,\varepsilon)) \leq C\varepsilon^Q. \tag{2.17}$$

Define N_ε to be the maximal number of disjoints balls of radius ε included in $B(p,r)$, and consider such a family $B(q_i,\varepsilon)$, $i=,\dots,N_\varepsilon$, of disjoints balls. By (2.17),

$$\frac{1}{C}\varepsilon^Q N_\varepsilon \leq \mathrm{vol}_\omega(B(p,r)) \quad \Rightarrow \quad N_\varepsilon \leq C\varepsilon^{-Q}\mathrm{vol}_\omega(B(p,r)).$$

On the other hand the union $\bigcup_i B(q_i,2\varepsilon)$ covers $B(p,r)$, and by Theorem 2.4 every ball $B(q_i,2\varepsilon)$ is of diameter $\geq \frac{4}{C}\varepsilon$ if ε is small enough. This implies

$$\mathscr{S}^Q(B(p,r)) \leq \liminf_{\varepsilon\to 0} N_\varepsilon \left(\frac{4\varepsilon}{C}\right)^Q < \infty.$$

Therefore $\dim_{\mathscr{H}} B(p,r) \leq Q$.

Conversely, let $\bigcup_i B(q_i,r_i)$ be a covering of $B(p,r)$ with balls of diameter not greater than ε. If ε is small enough, every r_i is smaller than ε_0 and there holds

$$\mathrm{vol}_\omega(B(p,r)) \leq \sum_i \mathrm{vol}_\omega(B(q_i,r_i)) \leq C \sum_i r_i^Q.$$

As a consequence, we have $\mathscr{S}^Q(B(p,r)) \geq \mathrm{vol}_\omega(B(p,r))/C$, which in turn implies $\dim_{\mathscr{H}} B(p,r) \geq Q$. This ends the proof. $\square$

When (M,d) is not equiregular, the Hausdorff dimension of balls centered at singular points behaves in a different way. Let us show it on an example.

Consider the Martinet space (Example 2.5), that is, $\mathbb{R}^3$ endowed with the sub-Riemannian distance associated with the vector fields

$$X_1 = \partial_x \quad \text{and} \quad X_2 = \partial_y + \frac{x^2}{2}\partial_z.$$

A point $q = (x,y,z)$ is regular if $x \neq 0$ and in this case $\sum_i w_i(q) = 4$, otherwise it is singular and $\sum_i w_i(q) = 5$.

Lemma 2.4 *Let p a point in the Martinet space.*

- *If p is regular, then $\dim_{\mathscr{H}} B(p,r) = 4$, and $\mathscr{H}^4(B(p,r))$ is finite.*
- *If p is singular, then $\dim_{\mathscr{H}} B(p,r) = 4$, but $\mathscr{H}^4(B(p,r))$ is not finite.*

Proof When p is regular, the result is a direct consequence of Theorem 2.8. Let us consider a singular point p and a radius $r > 0$. Since regular points form an open set, $B(p,r)$ contains small balls centered at regular points, and thus $\dim_{\mathscr{H}} B(p,r) \geq 4$.

Let us apply Corollary 2.3. We choose as a volume form $\omega = dx \wedge dy \wedge dz$, the associated measure vol_ω being the Lebesgue measure μ on $\mathbb{R}^3$. The function $f_{q,\varepsilon}$

takes only two values, $f_{q,\varepsilon}(\mathbf{X}) = |x|\varepsilon^4$ if $\mathbf{X} = (X_1, X_2, X_{[1,2]})$ and $f_{q,\varepsilon}(\mathbf{X}) = \varepsilon^5$ if $\mathbf{X} = (X_1, X_2, X_{[1,[1,2]]})$. Thus, for $q = (x, y, z)$ close enough to p and for $\varepsilon > 0$ small enough,

$$\frac{1}{C}\varepsilon^4 \max(|x|, \varepsilon) \leq \mu(B(q, \varepsilon)) \leq C\varepsilon^4 \max(|x|, \varepsilon). \qquad (2.18)$$

We proceed as in the proof of Theorem 2.8. Define N_ε to be the maximal number of disjoints balls of radius ε included in $B(p, r)$, and consider such a family $B(q_i, \varepsilon)$, $i =, \dots, N_\varepsilon$, of disjoints balls, with $q_i = (x_i, y_i, z_i)$. Notice that the first coordinate x is of nonholonomic order ≤ 1 at any point. This implies that there exists a constant $C' > 0$ such that

$$B(q_i, \varepsilon) \subset B(p, r) \cap \{q = (x, y, z) \ : \ |x - x_i| \leq C'\varepsilon\}.$$

As a consequence, for an integer k, every ball $B(q_i, \varepsilon)$ such that $(k-1)\varepsilon \leq |x_i| < k\varepsilon$ is included in the set $B(p, r) \cap \{q = (x, y, z) \ : \ |x| \in ((k - 1 - C')\varepsilon, (k + C')\varepsilon]\}$. The volume of the latter set is smaller than $C''\varepsilon$, where C'' is a constant (depending neither on k nor ε). Then it results from (2.18) that

$$\frac{1}{C}N_\varepsilon(k)k\varepsilon^5 \leq C''\varepsilon,$$

where $N_\varepsilon(k)$ is the number of points q_i such that $(k-1)\varepsilon \leq |x_i| < k\varepsilon$. The Ball-Box Theorem implies that $N_\varepsilon(k) = 0$ when $k > C'r/\varepsilon$, and hence

$$N_\varepsilon = \sum_{k=1}^{\lceil C'r/\varepsilon \rceil} N_\varepsilon(k) \leq \frac{\text{const}}{\varepsilon^4} \sum_{k=1}^{\lceil C'r/\varepsilon \rceil} \frac{1}{k} \leq \frac{\text{const}}{\varepsilon^4} \log\left(\frac{1}{\varepsilon}\right),$$

where $\lceil t \rceil$ denotes the integer part of a number t. Now the union $\bigcup_i B(q_i, 2\varepsilon)$ covers $B(p, r)$ and every ball $B(q_i, 2\varepsilon)$ is of diameter $\geq \frac{4}{C}\varepsilon$ if ε is small enough. This implies that, for any real number $s > 4$,

$$\mathscr{S}^s(B(p, r)) \leq \lim_{\varepsilon \to 0} \left(\frac{4\varepsilon}{C}\right)^s N_\varepsilon \leq \lim_{\varepsilon \to 0} \text{const}\, \varepsilon^{s-4} \log\left(\frac{1}{\varepsilon}\right) = 0.$$

Consequently $\dim_{\mathscr{H}} B(p, r) \leq 4$, and hence $\dim_{\mathscr{H}} B(p, r) = 4$, since the converse inequality holds.

We are left to show that $\mathscr{H}^4(B(p, r))$, or equivalently $\mathscr{S}^4(B(p, r))$, is not finite. Let $\bigcup_i B(q_i, r_i)$ be a covering of $B(p, r)$ with balls of diameter not greater than ε. For an integer $k \geq 1$, denote by $\mathscr{I}_k$ the set of indices such that $\bigcup_{i \in \mathscr{I}_k} B(q_i, r_i)$ is a covering of the set $B(p, r) \cap \{q = (x, y, z) \ : \ |x| \in ((k - 1)\varepsilon, k\varepsilon]\}$. Thus

$$\sum_{i \in \mathscr{I}_k} \mu(B(q_i, r_i)) \geq \text{const } \varepsilon.$$

On the other hand $i \in \mathscr{I}_k$ implies $\mu(B(q_i, r_i)) \leq \text{const } r_i^4 k \varepsilon$, and so

$$\sum_{i \in \mathscr{I}_k} r_i^4 \geq \frac{\text{const}}{k}.$$

Summing up over k, we obtain, for a small enough ε,

$$\sum_i r_i^4 \geq \text{const } \log\left(\frac{1}{\varepsilon}\right).$$

As a consequence, $\mathscr{S}_\varepsilon^4(B(p, r)) \geq \text{const } \log(\frac{1}{\varepsilon})$, and so $\mathscr{S}_\varepsilon^4(B(p, r)) = \infty$. This ends the proof. $\qquad\square$

Note that after the writing of these notes, new results on Hausdorff measures and dimensions in Carnot-Carathéodory spaces have been published (see [13]).

2.4 Desingularization

We have seen in the previous sections that the key feature of regular points is uniformity:

- uniformity of the flag (2.2);
- uniformity w.r.t. p of the convergence $(\lambda(M, d), p) \to C_p M$ (as explained by Bellaïche [4, Sect. 8], this uniformity is responsible for the group structure of the metric tangent space);
- uniformity of distance estimates (see Sect. 2.2.2).

All these uniformity properties are lost at singular points. We can however recover some of these properties by a process of desingularization.

2.4.1 Lifting of a Nonholonomic System

In order to desingularize a mathematical structure, the usual way is to consider a singularity as the projection of a regular object. We will then try to construct a manifold $\widetilde{M} = M \times \mathbb{R}^k$ and vector fields $\xi_1, \ldots, \xi_m$ such that:

- for $i = 1, \ldots, m$, X_i is the pushforward $\pi_* \xi_i$ of ξ_i by the canonical projection $\pi : \widetilde{M} \to M$, that is,

$$d\pi_{\widetilde{p}}(\xi(\widetilde{p})) = X_i(\pi(\widetilde{p})) \quad \text{for every } \widetilde{p} \in \widetilde{M};$$

when it is the case, we say that $(\xi_1, \ldots, \xi_m)$ is a *lifting of* $(X_1, \ldots, X_m)$ *to* $\widetilde{M}$;
- $\widetilde{M}$ equipped with $\xi_1, \ldots, \xi_m$ is equiregular (see Definition 2.10).

The algebraic structure of the metric tangent space provides a good indication on how to obtain such a lifting. Indeed, let us consider the particular case of vector fields $\widehat{X}_1, \ldots, \widehat{X}_m$ on $\mathbb{R}^n$ which are a nilpotent approximation of $X_1, \ldots, X_m$ at a singular point p. Keeping the notations and definitions of Sect. 2.3.1, we have the following diagram between Carnot-Carathéodory spaces,

$$
\begin{array}{c}
(G_p, d_{G_p}) \\
\pi \downarrow \qquad \phi_p \searrow \\
\qquad\qquad \overset{\psi_p}{} \\
(G_p/H_p, \overline{d}) \overset{}{\longrightarrow} (\mathbb{R}^n, \widehat{d})
\end{array}
$$

Thus up to an isomorphism $(\mathbb{R}^n, \widehat{d})$ is the projection of (G_p, d_{G_p}), which is an equiregular Carnot-Carathéodory space since the sub-Riemannian metric on G_p is right-invariant. Recall now that $\widehat{\xi}_1, \ldots, \widehat{\xi}_m$ (resp. $\overline{\xi}_1, \ldots, \overline{\xi}_m$) are mapped to $\widehat{X}_1, \ldots, \widehat{X}_m$ by ϕ_p (resp. ψ_p). Working in a system of coordinates, we identify G_p/H_p with $\mathbb{R}^n$ and $\overline{\xi}_i$ with $\widehat{X}_i$. These coordinates on $\mathbb{R}^n \simeq G_p/H_p$, denoted by x, induce coordinates $(x, y) \in \mathbb{R}^N$ on G_p for which we have

$$\widehat{\xi}_i(x, y) = \widehat{X}_i(x) + \sum_{j=n+1}^{N} b_{ij}(x, y)\partial_{y_j}. \tag{2.19}$$

Hence $\mathbb{R}^N$ equipped with $\widehat{\xi}_1, \ldots, \widehat{\xi}_m$ is an equiregular lifting of $\mathbb{R}^n$ equipped with $\widehat{X}_1, \ldots, \widehat{X}_m$.

We will use this idea to desingularize the original space (M, d). Choose for x privileged coordinates at p, so that

$$X_i(x) = \widehat{X}_i(x) + R_i(x) \qquad \text{with } \mathrm{ord}_p R_i \geq 0.$$

Set $\widetilde{M} = M \times \mathbb{R}^{N-n}$, and in local coordinates (x, y) on $\widetilde{M}$, define vector fields on a neighbourhood of $(p, 0)$ by

$$\xi_i(x, y) = X_i(x) + \sum_{j=n+1}^{N} b_{ij}(x, y)\partial_{y_j},$$

with the same functions b_{ij} as in (2.19). These vector fields realize locally a lifting of the vector fields $X_1, \ldots, X_m$ to $\widetilde{M}$: denoting by $\pi : \widetilde{M} \to M$ the canonical projection, we have $X_i = \pi_* \xi_i$ for $i = 1, \ldots, m$.

We define in this way a nonholonomic system on an open set $\widetilde{U} \subset \widetilde{M}$ whose nilpotent approximation at $(p, 0)$ is by construction $(\widehat{\xi}_1, \ldots, \widehat{\xi}_m)$. Unfortunately, $(p, 0)$ can be itself a singular point. Indeed, a point can be singular for a system and regular for the nilpotent approximation taken at this point.

Example 2.14 Take the vector fields $X_1 = \partial_{x_1}$, $X_2 = \partial_{x_2} + x_1 \partial_{x_3} + x_1^2 \partial_{x_4}$ and $X_3 = \partial_{x_5} + x_1^{100} \partial_{x_4}$ on $\mathbb{R}^5$. The origin 0 is a singular point. However the nilpotent approximation at 0 is $\widehat{X}_1 = X_1$, $\widehat{X}_2 = X_2$, $\widehat{X}_3 = \partial_{x_5}$, for which 0 is not singular.

To avoid this difficulty, we need a group bigger than G_p, namely the free nilpotent group N_r of step r with m generators. Let us first recall some general facts on free Lie algebras (see [5] for more details).

Let $\mathscr{L} = \mathscr{L}(1, \ldots, m)$ be the free Lie algebra generated by $\{1, \ldots, m\}$. We use $\mathscr{L}^s$ to denote the subspace generated by elements of $\mathscr{L}$ of length not greater than s, and $\widetilde{n}_s$ to denote the dimension of $\mathscr{L}^s$.

Definition 2.13 Let $\xi_1, \ldots, \xi_m$ be m vector fields on a manifold $\widetilde{M}$, and r be a positive integer. The Lie algebra $Lie(\xi_1, \ldots, \xi_m)$ is said to be *free up to step r* if, for every $x \in \widetilde{M}$, the elements $n_1(x), \ldots, n_r(x)$ of the growth vector are equal to $\widetilde{n}_1, \ldots, \widetilde{n}_r$.

Remark 2.5 Consider a manifold $\widetilde{M}$ of dimension $\widetilde{n}_r$. If $Lie(\xi_1, \ldots, \xi_m)$ is free up to step r, then every point in $\widetilde{M}$ is regular. It is in particular the case when the degree of nonholonomy of $(\xi_1, \ldots, \xi_m)$ equals r at every point of $\widetilde{M}$.

Let $\mathscr{L}^{(s)}$ be the subspace of $\mathscr{L}$ generated by elements of length equal to s. Then $\mathfrak{n}_r = \mathscr{L}/\mathscr{L}^{(r+1)}$ is nilpotent of step r and is called *the free nilpotent Lie algebra of step r* generated by $\{1, \ldots, m\}$. The corresponding simply connected Lie group $N_r = \exp(\mathfrak{n}_r)$ is *the free nilpotent group of step r*. It is a Carnot group, and the generators $\alpha_1, \ldots, \alpha_m$ ($\alpha_i = i \mod \mathscr{L}^{(r+1)}$) of $\mathfrak{n}_r$ define on N_r a right-invariant sub-Riemannian distance d_N.

Now, the group N_r defines a left action on $\mathbb{R}^n$: given $g \in N_r$, we have $g = \exp(I)$ where $I \in \mathfrak{n}_r$, and the action is defined by

$$g : q \in \mathbb{R}^n \mapsto g \cdot q = \exp(\widehat{X}_I)(q),$$

where $\widehat{X}_I \in Lie(\widehat{X}_1, \ldots, \widehat{X}_m)$ is obtained by plugging in $\widehat{X}_i$, $i = 1, \ldots, m$, for the corresponding letter i in I. Denoting by K the isotropy subgroup of 0 for this action, we obtain that $(\mathbb{R}^n, \widehat{d})$ is isometric to N_r/K endowed with the restriction of the distance d_N.

Reasoning as we did above with (G_p, d_{G_p}), we are able to lift locally the vector fields $X_1, \ldots, X_m$ on M to vector fields on $M \times \mathbb{R}^{\widetilde{n}_r - n}$ having $\alpha_1, \ldots, \alpha_m$ for nilpotent approximation at $(p, 0)$. Moreover $(p, 0)$ is a regular point for the associated nonholonomic system in $M \times \mathbb{R}^{\widetilde{n}_r - n}$ since N_r is free up to step r. We obtain in this way a result of desingularization.

Lemma 2.5 *Let p be a point in M, r the degree of nonholonomy at p, and $\widetilde{M} = M \times \mathbb{R}^{\widetilde{n}_r - n}$. Then there exist a neighbourhood $\widetilde{U} \subset \widetilde{M}$ of $(p, 0)$, a neighbourhood $U \subset M$ of p with $U \times \{0\} \subset \widetilde{U}$, local coordinates (x, y) on $\widetilde{U}$, and smooth vector fields on $\widetilde{U}$,*

$$\xi_i(x, y) = X_i(x) + \sum_{j=n+1}^{N} b_{ij}(x, y)\partial_{y_j}, \quad i = 1, \ldots, m, \tag{2.20}$$

such that:

- *$\xi_1, \ldots, \xi_m$ satisfy Chow's Condition and have r for degree of nonholonomy everywhere (so the Lie algebra they generate is free up to step r);*
- *every $\widetilde{q}$ in $\widetilde{U}$ is regular;*
- *denoting by $\pi : \widetilde{M} \to M$ the canonical projection, and by $\widetilde{d}$ the sub-Riemannian distance defined by $\xi_1, \ldots, \xi_m$ on $\widetilde{U}$, we have $\pi_*\xi_i = X_i$, and for $q \in U$ and $\varepsilon > 0$ small enough,*

$$B(q, \varepsilon) = \pi\left(B^{\widetilde{d}}\big((q, 0), \varepsilon\big)\right),$$

or, equivalently,

$$d(q_1, q_2) = \inf_{\widetilde{q}_2 \in \pi^{-1}(q_2)} \widetilde{d}\big((q_1, 0), \widetilde{q}_2\big).$$

Proof The only things that remain to prove are the last equalities. Let $(x(\cdot), y(\cdot))$ be a trajectory of the nonholonomic system in $\widetilde{U}$ defined by $\xi_1, \ldots, \xi_m$. Then, for every control $u(\cdot)$ associated with the trajectory,

$$(\dot{x}(t), \dot{y}(t)) = \sum_{i=1}^{m} u_i(t)\xi_i(x, y).$$

It follows from (2.20) that $x(\cdot)$ is a trajectory in U of the system defined by $X_1, \ldots, X_m$, which is associated with the same controls $u(\cdot)$, so that

$$\text{length}\big(x(\cdot)\big) = \text{length}\big((x, y)(\cdot)\big).$$

The relation between d and $\widetilde{d}$ follows. $\qquad\square$

Remark 2.6 The lemma still holds if we replace r by any integer greater than the degree of nonholonomy at p.

Thus any Carnot-Carathéodory space (M, d) is locally the projection of an equiregular Carnot-Carathéodory space $(\widetilde{M}, \widetilde{d})$. This projection preserves the trajectories, the minimizers, and the distance.

Example 2.15 (Martinet case) Consider the vector fields of the Martinet case (see Example 2.5), defined on $\mathbb{R}^3$ by:

$$X_1 = \partial_x \quad \text{and} \quad X_2 = \partial_y + \frac{x^2}{2}\partial_z.$$

Let $\pi : \mathbb{R}^4 \to \mathbb{R}^3$ be the projection with respect to the last coordinates, $\pi(x, y, z, w) = (x, y, z)$. Then X_1 and X_2 are the projections of the vector fields defining the Engel group $\mathbb{E}^4$ (see Example 2.12),

$$\xi_1 = \partial_x \quad \text{and} \quad \xi_2 = \partial_y + \frac{x^2}{2}\partial_z + x\partial_w,$$

that is $\pi_*\xi_i = X_i$. Thus, for every pair of points $q_1, q_2 \in \mathbb{R}^3$,

$$d_{\mathrm{Mart}}(q_1, q_2) = \inf_{w\in\mathbb{R}} d_{\mathbb{E}^4}\big((q_1, 0), (q_2, w)\big),$$

where d_{Mart} and $d_{\mathbb{E}^4}$ are the sub-Riemannian distance in respectively the Martinet space and the Engel group.

Example 2.16 (Grušin plane) Consider the vector fields

$$X_1 = \partial_x, \qquad X_2 = x\partial_y,$$

on $\mathbb{R}^2$, which define the Grušin plane (see Example 2.11). Let $\pi : \mathbb{R}^3 \to \mathbb{R}^2$ be the projection with respect to the last coordinates, $\pi(x, y, z) = (x, y)$. Then $X_1 = \pi_*\xi_1$ and $X_2 = \pi_*\xi_2$, where

$$\xi_1 = \partial_x \quad \text{and} \quad \xi_2 = \partial_z + x\partial_y,$$

are, up to a change of coordinates, the vector fields defining the Heisenberg case (see Example 2.3).

2.4.2 Desingularization Procedure

Lemma 2.5 states the existence of a desingularized lifting of the vector fields $X_1, \ldots, X_m$. Our aim now is to give an effective construction involving only explicit (and purely algebraic) changes of coordinates and intermediate constructions. We will also obtain, as a byproduct, a nilpotent approximation of the lifting in a "canonical form". We begin with the presentation of this canonical form.

P. Hall Basis and Canonical Form

Recall that we use $\mathscr{L} = \mathscr{L}(1, \ldots, m)$ to denote the free Lie algebra generated by the elements $\{1, \ldots, m\}$. The *P. Hall basis* of $\mathscr{L}$ is a subset $\mathscr{H} = \{I_j\}_{j \in \mathbb{N}}$ of $\mathscr{L}$, endowed with a total order $\prec$, that satisfies the following:

(H1) if $|I_i| < |I_j|$, then $I_i \prec I_j$;

(H2) $\{1, \ldots, m\} \subset \mathscr{H}$, and we impose that $1 \prec 2 \prec \cdots \prec m$;

(H3) every element of length 2 in $\mathscr{H}$ is in the form $[I_i, I_j]$ with $I_i, I_j \in \{1, \ldots, m\}$ and $I_i \prec I_j$;

(H4) an element $I_k \in \mathscr{L}$ of length greater than 3 belongs to $\mathscr{H}$ if $I_k = [I_{k_1}, [I_{k_2}, I_{k_3}]]$ with $I_{k_1}, I_{k_2}, I_{k_3} \in \mathscr{H}$, and (i) $[I_{k_2}, I_{k_3}]$ belongs to $\mathscr{H}$, (ii) $I_{k_2} \prec I_{k_3}$, (iii) $I_{k_2} \prec I_{k_1}$ or $I_{k_2} = I_{k_1}$, and (iv) $I_{k_1} \prec [I_{k_2}, I_{k_3}]$.

The elements of $\mathscr{H}$ form a basis of $\mathscr{L}$, and $\prec$ defines a strict and total order over the set $\mathscr{H}$. In the sequel, we use I_k to denote the kth element of $\mathscr{H}$ with respect to that order. Let $\mathscr{H}^s$ be the subset of $\mathscr{H}$ of all the elements of length not greater than s. The elements of $\mathscr{H}^s$ form a basis of $\mathscr{L}^s$ and $\mathrm{Card}(\mathscr{H}^s) = \tilde{n}_s$.

By (H1)−(H4), every element $I_j \in \mathscr{H}$ can be expanded in a unique way as

$$I_j = [I_{k_1}, [I_{k_2}, \ldots, [I_{k_i}, I_k] \ldots]], \tag{2.21}$$

with $k_1 \geq \cdots \geq k_i$, $k_i < k$, and $k \in \{1, \ldots, \tilde{n}_1\}$. We set $\phi(I_j) = k$. For $I_j \in \mathscr{H}^r$, the expansion (2.21) also associates with $I_j \in \mathscr{H}$ a sequence $\alpha_j = (\alpha_j^1, \ldots, \alpha_j^{\tilde{n}_r})$ in $\mathbb{Z}^{\tilde{n}_r}$ where α_j^ℓ is the number of occurrences of ℓ among $k_1, \ldots, k_i$. By construction, one has $\alpha_j^\ell = 0$ for $\ell \geq j$, and $\alpha_j = (0, \ldots, 0)$ for $1 \leq j \leq \tilde{n}_1$.

Let us give now the construction of a canonical form for free nilpotent Lie algebras of step r. The construction takes place in $\mathbb{R}^{\tilde{n}_r}$, where r is a positive integer and $\tilde{n}_r$ the dimension of $\mathscr{L}^r$. Let $I_1, \ldots, I_{\tilde{n}_r}$ be the elements of $\mathscr{H}^r$ sorted by increasing order with respect to $\prec$. We define the monomials $P_j(x)$, $j = 1, \ldots, \tilde{n}_r$, by

$$P_j(x) = \frac{x_1^{\alpha_j^1} \ldots x_{j-1}^{\alpha_j^{j-1}}}{\alpha_j^1! \ldots \alpha_j^{j-1}!}, \tag{2.22}$$

where α_j is the sequence associated with I_j.

Definition 2.14 Let $X_1, \ldots, X_m$ be m vector fields on an open subset Ω of $\mathbb{R}^{\tilde{n}_r}$ and x be a local system of coordinates on Ω. The family $(X_1, \ldots, X_m)$ is said to be in *canonical form in the coordinates x* if one has

$$x_* X_1 = \partial_{x_1},$$
$$x_* X_i = \partial_{x_i} + \sum_{\substack{2 \leq |I_j| \leq r \\ \phi(I_j) = i}} P_j(x) \partial_{x_j}, \quad i = 2, \ldots, m.$$

In this case, $Lie(X_1, \ldots, X_m)$ is a free nilpotent Lie algebra of step r, and one has

$$X_{I_j}(0) = \partial_{x_j}, \quad \text{for } I_j \in \mathcal{H}^r,$$

where the vector field $X_{I_j} \in Lie(X_1, \ldots, X_m)$ is obtained by plugging in X_i, $i = 1, \ldots, m$, for the corresponding letter i in I_j (see [10, 11] for a complete development on this canonical form).

Reasoning by induction on the coefficients P_j, one can show that the nonholonomic system $\dot{x} = \sum_{i=1}^{m} u_i X_i(x)$ associated with a canonical form may be written component by component as,

$$\dot{x}_j = \frac{1}{k!} x_{j_1}^k \dot{x}_{j_2}, \quad \text{if } I_j = \mathrm{ad}_{I_{j_1}}^k I_{j_2}, \tag{2.23}$$

where $\mathrm{ad}_{I_{j_1}}^k I_{j_2} = \underbrace{[I_{j_1}, [I_{j_1}, \ldots, [I_{j_1}, I_{j_2}]}_{k \text{ times}}$, with $I_{j_2} = [I_{j_3}, I_{j_4}]$ and $I_{j_3} \prec I_{j_1}$.

Remark 2.7 Note that the canonical coordinates of the second kind (see Sect. B.1) allows one to obtain canonical forms. Indeed, assume that $X_1, \ldots, X_m$ generate a free nilpotent Lie algebra of step r. Then $(X_{I_1}, \ldots, X_{I_{\tilde{n}_r}})$ form an adapted basis at every point and $(X_1, \ldots, X_m)$ is in canonical form in the canonical coordinates of the second kind associated with that basis (see [25]).

Construction of a Desingularized Lifting

As we are concerned here with algorithmic issues, we work on an open subset Ω of $\mathbb{R}^n$ rather than on a general manifold. We denote by x the canonical coordinates on Ω.

Consider m vector fields $X_1, \ldots, X_m$ on Ω satisfying Chow's Condition, and fix a point $p \in \Omega$. Choose also an integer r greater than or equal to the degree of nonholonomy at p, and a n-tuple $\mathcal{J} = (I_1, \ldots, I_n)$ of elements of $\mathcal{H}^r$ such that the vectors $X_{I_1}(p), \ldots, X_{I_n}(p)$ form a basis of $\mathbb{R}^n$. We define an open domain $\mathcal{V}_{\mathcal{J}} \subset \Omega$ containing p by

$$\mathcal{V}_{\mathcal{J}} = \{\, x \in \Omega \ : \ \det(X_{I_1}(x), \ldots, X_{I_n}(x)) \neq 0 \,\}.$$

We are going to construct a family of m vector fields $(\xi_1, \ldots, \xi_m)$ defined on $\mathcal{V}_{\mathcal{J}} \times \mathbb{R}^{\tilde{n}_r - n}$, which is a lifting of $(X_1, \ldots, X_m)$, and which generates a free up to step r Lie algebra. At the same time, we will give a nilpotent approximation of $(\xi_1, \ldots, \xi_m)$ at $\tilde{p} = (p, 0) \in \mathcal{V}_{\mathcal{J}} \times \mathbb{R}^{\tilde{n}_r - n}$ in canonical form.

Define

$$\mathcal{J}^s = \{I_j \in \mathcal{J}, \text{ with } |I_j| = s\}, \qquad \text{for } s \geq 1,$$
$$\mathcal{G}^s = \mathcal{H}^s \setminus \mathcal{H}^{s-1}, \qquad \text{for } s \geq 2.$$

We denote by k_s the cardinal of $\mathcal{J}^s$, and by $\tilde{k}_s$ the cardinal of $\mathcal{G}^s$. We also define $\tilde{w}_1, \ldots, \tilde{w}_{\tilde{n}_r}$, called the *free weights of step r*, by $\tilde{w}_j = s$ if $\tilde{n}_{s-1} < j \leq \tilde{n}_s$. We are now ready to describe our procedure.

Desingularization Algorithm

Initialization step

(i) Set $\mathcal{V}^1 := \mathcal{V}_{\mathcal{J}} \times \mathbb{R}^{\tilde{k}_1 - k_1}$, $p^1 := (p, 0) \in \mathcal{V}^1$, $\mathcal{K}^1 := \mathcal{H}^1 \cup (\mathcal{J} \setminus \mathcal{J}^1)$.

   Denote by v^1 the points in $\mathbb{R}^{\tilde{k}_1 - k_1}$.

(ii) Define the vector fields $\xi_1^1, \ldots, \xi_m^1$ on $\mathcal{V}^1$ by:

$$\forall (x, v^1) \in \mathcal{V}^1, \qquad \xi_i^1(x, v^1) := X_i(x) + \begin{cases} 0, & \text{for } i \in \mathcal{J}^1, \\ \partial_{v_i^1}, & \text{for } i \in \mathcal{G}^1 \setminus \mathcal{J}^1. \end{cases}$$

(iii) Compute the affine coordinates y^1 on $\mathcal{V}^1$ satisfying

$$\partial_{y_j^1} = \xi_{I_j}^1(p^1) \text{ for } I_j \in \mathcal{K}^1, \quad \text{and} \quad y^1(p^1) = 0.$$

(iv) Construct the coordinates z^1 on $\mathcal{V}^1$ by

$$z_j^1 := y_j^1, \quad \text{for } j \in \mathcal{H}^1,$$
$$z_j^1 := y_j^1 - \sum_{k=1}^{\tilde{n}_1} (\xi_k^1 \cdot y_k^1)(y^1)_{|y^1=0} \; y_k^1, \quad \text{for } I_j \in \mathcal{K}^1 \setminus \mathcal{H}^1,$$

   where I_j denotes the jth element in $\mathcal{K}^1$.

Iteration steps

For $s = 2, \ldots, r$:

(i) Set $\mathcal{V}^s := \mathcal{V}^{s-1} \times \mathbb{R}^{\tilde{k}_s - k_s}$, $p^s := (p, 0) \in \mathcal{V}^s$, and $\mathcal{K}^s := \mathcal{K}^{s-1} \cup (\mathcal{G}^s \setminus \mathcal{J}^s)$.

   Denote by v^s the points in $\mathbb{R}^{\tilde{k}_s - k_s}$.

(ii) Define the vector fields $\xi_1^s, \ldots, \xi_m^s$ on $\mathcal{V}^s$ by:

$$\forall (z^{s-1}, v^s) \in \mathcal{V}^s, \qquad \xi_i^s(z^{s-1}, v^s) = \xi_i^{s-1}(z^{s-1}) + \sum_{\substack{I_k \in \mathcal{G}^s \setminus \mathcal{J}^s \\ \phi(I_k)=i}} P_k(z^{s-1}) \partial_{v_k^s}.$$

(iii) Compute the linear change of coordinates $(z^{s-1}, v^s) \mapsto y^s$ on $\mathscr{V}^s$ by

$$\partial_{y_j^s} = \xi_{I_j}^s(p^s) \quad \text{for every } I_j \in \mathscr{K}^s \,.$$

(iv) Construct the coordinates $\tilde{z}^s$ on $\mathscr{V}^s$ by the following recursive formulas:

- for $I_j \in \mathscr{H}^{\wp s}$,

$$\tilde{z}_j^s := y_j^s - \sum_{k=2}^{|I_j|-1} r_k(y_1^s, \ldots, y_{j-1}^s), \tag{2.24}$$

where, for $k = 2, \ldots, |I_j| - 1$,

$$r_k(y_1^s, \ldots, y_{j-1}^s)$$
$$= \sum_{\substack{|\beta|=k \\ \widetilde{w}(\beta)<|I_j|}} \Big[(\xi_{I_1}^s)^{\beta_1} \cdots (\xi_{I_{j-1}}^s)^{\beta_{j-1}} \cdot (y_j^s - \sum_{q=2}^{k-1} r_q) \Big](y^s)_{|y^s=0} \frac{(y_1^s)^{\beta_1}}{\beta_1!} \cdots \frac{(y_{j-1}^s)^{\beta_{j-1}}}{\beta_{j-1}!} \,;$$

- for $I_j \in \mathscr{K}^s \setminus \mathscr{H}^{\wp s}$,

$$\tilde{z}_j^s := y_j^s - \sum_{k=2}^{s} r_k(y_1^s, \ldots, y_{\tilde{n}_s}^s), \tag{2.25}$$

where, for $k = 2, \ldots, s$,

$$r_k(y_1^s, \ldots, y_{\tilde{n}_s}^s)$$
$$= \sum_{\substack{|\beta|=k \\ \widetilde{w}(\beta)\leq s}} \Big[(\xi_{I_1}^s)^{\beta_1} \cdots (\xi_{I_{\tilde{n}_s}}^s)^{\beta_{\tilde{n}_s}} \cdot (y_j^s - \sum_{q=2}^{s} r_q) \Big](y^s)_{|y^s=0} \frac{(y_1^s)^{\beta_1}}{\beta_1!} \cdots \frac{(y_{j-1}^s)^{\beta_{\tilde{n}_s}}}{\beta_{\tilde{n}_s}!} \,.$$

(v) Construct the coordinates z^s as:

$$\begin{cases} z_j^s := \tilde{z}_j^s + \Psi_j^s(\tilde{z}_1^s, \ldots, \tilde{z}_{j-1}^s) & \text{for } j = 1, \ldots, \tilde{n}_s, \\ z_j^s := \tilde{z}_j^s & \text{for } j > \tilde{n}_s, \end{cases}$$

where the functions Ψ_j^s are determined by the following properties:

- for $j = 1, \ldots, \tilde{n}_s$, Ψ_j^s is an homogeneous polynomial of weighted degree equal to $\widetilde{w}_j$, the weight of a coordinate $\tilde{z}_k^s$ being $\widetilde{w}_k$;
- denoting by $\mathrm{ord}_{p^s}^s(\cdot)$ the nonholonomic order at p^s defined by $(\xi_1^s, \ldots, \xi_m^s)$, one has

$$\xi_i^s \cdot z_j^s = \delta_{i,\phi(I_j)} P_j(z_1^s, \ldots, z_{j-1}^s) + R_{i,j}(z^s), \quad j = 1, \ldots, \widetilde{n}_s, \qquad (2.26)$$

where $\mathrm{ord}_{p^s}^s(R_{i,j}) \geq \widetilde{w}_j$ and $\delta_{i,k}$ denotes the Kronecker symbol (note that by construction $\mathrm{ord}_{p^s}^s(P_j) = \widetilde{w}_j - 1$).

Remark 2.8 The polynomials $(\Psi_j^s)_{j=1,\ldots,\widetilde{n}_s}$ are computed by identification. Indeed, the first property above implies that Ψ_j^s is a finite sum of the form,

$$\Psi_j^s(\widetilde{z}^s) = \sum_{\widetilde{w}(\alpha)=\widetilde{w}_j} c_j^\alpha (\widetilde{z}_1^s)^{\alpha_1} \ldots (\widetilde{z}_{j-1}^s)^{\alpha_{j-1}}, \qquad (2.27)$$

where every c_j^α is a real number. One can then put this expression in (2.26), and so obtain by identification the scalar coefficients c_j^α, provided they exist. Actually it is possible to show that these coefficients do exist (see [6, Claim 9 of Sect. 3.4]).

Let us illustrate the previous remark with an example.

Example 2.17 Assume $m = 2$, $I_1 = 1$, $I_2 = 2$, and $I_3 = [1, 2]$, and consider the previous algorithm at the step $s = 2$. Item (iv) produces coordinates $\widetilde{z}$ and vector fields ξ_1, ξ_2 which are necessarily of the form:

$$\xi_1 = \left[1 + R_{1,1}(\widetilde{z})\right] \partial_{\widetilde{z}_1} + R_{1,2}(\widetilde{z})\partial_{\widetilde{z}_2} + [\alpha_1 \widetilde{z}_1 + \alpha_2 \widetilde{z}_2 + R_{1,3}(\widetilde{z})]\partial_{\widetilde{z}_3} + \cdots$$
$$\xi_2 = R_{2,1}(\widetilde{z})\partial_{\widetilde{z}_1} + [1 + R_{2,2}(\widetilde{z})]\partial_{\widetilde{z}_2} + [\beta_1 \widetilde{z}_1 + \beta_2 \widetilde{z}_2 + R_{2,3}(\widetilde{z})]\partial_{\widetilde{z}_3} + \cdots$$

where the real numbers $\alpha_1, \alpha_2, \beta_1, \beta_2$ verify $\beta_1 - \alpha_2 = 1$, and where every $R_{i,j}$ is a sum of homogeneous polynomials of weighted degree greater than $\widetilde{w}_j - 1$.

The coordinates z of item (v) satisfy $(z_1, z_2) = (\widetilde{z}_1, \widetilde{z}_2)$, $z_3 = \widetilde{z}_3 + \Psi_3(\widetilde{z})$, and $z_j = \widetilde{z}_j$ for $j > 3$. We look for Ψ_3 as a linear combination,

$$\Psi_3(\widetilde{z}) = a\widetilde{z}_1\widetilde{z}_2 + b\widetilde{z}_1^2 + c\widetilde{z}_2^2,$$

with a, b, c to be determined. The constraints (2.26) imply $\xi_1 \cdot z_3 = 0$ and $\xi_2 \cdot z_3 = z_1$ up to higher orders terms, that is,

$$(\alpha_1 + 2b)\widetilde{z}_1 + (\alpha_2 + a)\widetilde{z}_2 = 0,$$
$$(\beta_1 + a)\widetilde{z}_1 + (\beta_2 + 2c)\widetilde{z}_2 = z_1 = \widetilde{z}_1.$$

By identification, one gets $a = -\alpha_2$, $b = -\frac{\alpha_1}{2}$, $c = -\frac{\beta_2}{2}$ and $\beta_1 + a = 1$. Since $\beta_1 - \alpha_2 = 1$ the last equation is automatically verified. Then

$$\Psi_3(\widetilde{z}) = -\alpha_2\widetilde{z}_1\widetilde{z}_2 - \frac{\alpha_1}{2}\widetilde{z}_1^2 - \frac{\beta_2}{2}\widetilde{z}_2^2.$$

The outputs of this Desingularization Algorithm are the coordinates z^r defined on $\mathcal{V}_{\mathcal{J}} \times \mathbb{R}^{\widetilde{n}_r - n}$, and the vector fields $\xi_1^r, \ldots, \xi_m^r$ defined on this domain. Note that

z^r are global coordinates on $\mathscr{V}_{\mathscr{J}} \times \mathbb{R}^{\tilde{n}_r - n}$ since they are obtained from the original coordinates (x, v) by a triangular change of coordinates.

Theorem 2.9 *The outputs $\xi_i := \xi_i^r$, $i = 1, \ldots, m$, and $z := z^r$ of the Desingularization Algorithm satisfy the following properties:*

- *the vector fields $\xi_1, \ldots, \xi_m$ realize a lifting of $X_1, \ldots, X_m$ to $\mathscr{V}_{\mathscr{J}} \times \mathbb{R}^{\tilde{n}_r - n}$;*
- *$\xi_1, \ldots, \xi_m$ generate a Lie algebra which is free up to step r;*
- *$z = (z_1, \ldots, z_{\tilde{n}_r})$ is a system of privileged coordinates at $\tilde{p}$ for $(\xi_1, \ldots, \xi_m)$;*
- *the vector fields $(\widehat{\xi}_1, \ldots, \widehat{\xi}_m)$ defined in the coordinates z by the canonical form,*

$$\widehat{\xi}_i = \partial_{z_i} + \sum_{\substack{2 \le |I_j| \le \tilde{n}_r \\ \phi(I_j) = i}} P_j(z_1, \ldots, z_{j-1}) \partial_{z_j}, \quad \text{for } i = 1, \ldots, m, \tag{2.28}$$

form a nilpotent approximation of $(\xi_1, \ldots, \xi_m)$ at $\tilde{p}$.

We refer to [6, Sect. 3.4] for a proof of this result.

Remark 2.9 A very nice feature of the Desingularization Algorithm is that, if the original Lie algebra $Lie(X_1, \ldots, X_m)$ is nilpotent, then the lifting $(\xi_1, \ldots, \xi_m)$ generates also a nilpotent Lie algebra. In fact, $(\xi_1, \ldots, \xi_m)$ is equal to its nilpotent approximation $(\widehat{\xi}_1, \ldots, \widehat{\xi}_m)$ at $\tilde{p}$, and so is in the canonical form in the coordinates z, which implies in particular that $Lie(\xi_1, \ldots, \xi_m)$ is a free nilpotent algebra of step r. Note that in that case r may be chosen as the nilpotency step of $(X_1, \ldots, X_m)$.

Remark 2.10 As a consequence of the Remark 2.7, the coordinates z are the canonical coordinates of the second kind on $\mathscr{V}_{\mathscr{J}} \times \mathbb{R}^{\tilde{n}_r - n}$ associated with the basis $(\widehat{\xi}_{I_1}, \ldots, \widehat{\xi}_{I_{\tilde{n}_r}})$. Thus the algorithm above provides an algebraic construction of these canonical coordinates.

References

1. Agrachev, A., Barilari, D., Boscain, U.: On the Hausdorff volume in sub-Riemannian geometry. Calc. Var. Partial Differ. Equ. **43**(3–4), 355–388 (2011)
2. Agrachev, A.A., Sarychev, A.V.: Filtrations of a Lie algebra of vector fields and nilpotent approximations of control systems. Dokl. Akad. Nauk SSSR **285**, 777–781 (1987). (English transl.: Soviet Math. Dokl. **36**, 104–108 (1988))
3. Burago, D., Burago, Y., Ivanov, S.: A Course in Metric Geometry, vol. 33 of Graduate Studies in Mathematics. American Mathematical Society (2001)
4. Bellaïche, A.: The tangent space in sub-Riemannian geometry. In: Bellaïche, A., Risler, J.-J. (eds.) Sub-Riemannian Geometry, Progress in Mathematics. Birkhäuser, Switzerland (1996)
5. Bourbaki, N: Groupes et Algèbres de Lie. Hermann, Paris (1972)
6. Chitour, Y., Jean, F., Long, R.: A global steering method for nonholonomic systems. J. Differ. Equ. **254**, 1903–1956 (2013)
7. Falbel, E., Jean, F.: Measures of transverse paths in sub-Riemannian geometry. J. d'Analyse Math. **91**, 231–246 (2003)

8. Gabrielov, A.: Multiplicities of zeroes of polynomials on trajectories of polynomial vector fields and bounds on degree of nonholonomy. Math. Res. Lett. **2**, 437–451 (1995)
9. Gershkovich, V.Y.: Two-sided estimates of metrics generated by absolutely non-holonomic distributions on Riemannian manifolds. Sov. Math. Dokl. **30**, 506–510 (1984)
10. Grayson, M., Grossman, R.: Nilpotent lie algebras and vector fields. In: Grossman, R. (eds) Symbolic Computation: Applications to Scientific Computing, pp. 77–96. SIAM (1989)
11. Grayson, M., Grossman, R.: Models for free nilpotent lie algebra. J. Algebra **135**, 177–191 (1991)
12. Ghezzi, R., Jean, F.: A new class of $(\mathscr{H},)$-rectifiable subsets of metric spaces. Com. Pure Appl. Anal. **12**, 881–898 (2013)
13. Ghezzi, R., Jean, F.: Geometric control theory and sub-Riemannian geometry, vol. 5, chapter Hausdorff measures and dimensions in non equiregular sub-Riemannian manifolds. Springer INdAM Series, Berlin (2014)
14. Gabrielov, A., Jean, F., Risler, J.-J.: Multiplicity of polynomials on trajectories of polynomials vector fields in C^3. In: Pawłucki, W., Jakubczyk, B., Stasica, J. (eds.) Singularities Symposium—Łojasiewicz 70. vol. 44, pp. 109–121. Banach Center Publications, Warszawa (1998)
15. Gromov, M.: Carnot-Carathéodory spaces seen from within. In: Bellaïche, A., Risler, J.-J. (eds.) Sub-Riemannian Geometry, Progress in Mathematics. Birkhäuser, Switzerland (1996)
16. Jean, F.: Uniform estimation of sub-Riemannian balls. J. Dyn. Control Syst. **7**(4), 473–500 (2001)
17. Kupka, I.: Géométrie sous-riemannienne. In: Bourbaki, S. (eds.), vol. 817, June (1996)
18. Lee, J.: Introduction To Smooth Manifolds, vol. 218 of Graduate Texts in Mathematics. Springer, Berlin (2003)
19. Mitchell, J.: On Carnot-Carathéodory metrics. J. Differ. Geom. **21**, 35–45 (1985)
20. Margulis, G.A., Mostow, G.D.: Some remarks on the definition of tangent cones in a Carnot-Carathéodory space. J. d'Analyse Math. **80**, 299–317 (2000)
21. Montgomery, R.: A tour of sub-Riemannian geometries, their geodesics and applications. Math. Surv. Monogr. Am. Math. Soc. (2002)
22. Nagel, A., Stein, E.M., Wainger, S.: Metrics defined by vector fields. Acta Math. **155**, 103–147 (1985)
23. Rudin, W.: Real and complex analysis. McGraw-Hill, New York (1970)
24. Stefani, G.: On local controllability of a scalar-input system. In: Byrnes, L. (ed.) Theory and Application of Nonlinear Control System, pp. 167–179. North Holland, Amsterdam (1986)
25. Sussmann, H.J.: A general theorem on local controllability. SIAM J. Control Optim. **25**, 158–194 (1987)

Chapter 3
Nonholonomic Motion Planning

Abstract This last chapter addresses one of the most basic control issue: how to construct a control law steering a control system from a given state to another one? This problem is known as the *motion planning problem*. In the case of nonholonomic systems it can be solved exactly for specific classes of systems, in particular for *nilpotent systems* (Sect. 3.2). However for a general nonholonomic systems it is hopeless to look for exact solutions to the problem. Thus a solution must be thought as an algorithm which steers the system to an arbitrarily small neighbourhood of the goal. In this context a key notion is the one of approximation and we will see in Sect. 3.3 how the concepts introduced in the previous chapter allow to construct such an algorithm. We will also discuss two other methods in Sect. 3.4 and give an overview of the literature in Sect. 3.5.

Keywords Control theory · Nonholonomic systems · Motion planning · Nilpotent systems

As we are concerned here with algorithmic issues, we will not work in a general manifold but rather on an open subset of $\mathbb{R}^n$. Thus we consider a nonholonomic system

$$\dot{x} = \sum_{i=1}^{m} u_i X_i(x), \tag{3.1}$$

defined on an open subset Ω of $\mathbb{R}^n$.

3.1 Nonholonomic Motion Planning

Definition 3.1 The *motion planning problem* for (3.1) is defined as follows: for each pair of points $(x^{\text{initial}}, x^{\text{final}}) \in \Omega \times \Omega$, find a control $u(\cdot) \in L^1([0, T], \mathbb{R}^m)$ with $T > 0$ such that the corresponding trajectory of (3.1) starting from x^{initial} at $t = 0$ reaches x^{final} at $t = T$, that is,

$$\gamma(T; x^{\text{initial}}, u) = x^{\text{final}}.$$

© The Author(s) 2014

F. Jean, *Control of Nonholonomic Systems: From Sub-Riemannian Geometry to Motion Planning*, SpringerBriefs in Mathematics, DOI 10.1007/978-3-319-08690-3_3

We will always assume that Ω is connected and that (3.1) satisfies the Chow Condition. The existence of a solution to the motion planning problem is then guaranteed by Chow-Rashevsky's theorem (Theorem 1.1). Note that in practice the vector fields $X_1, \ldots, X_m$ generally belong to the analytic category, and in this case our assumptions are equivalent to the existence of a solution (see Remark 1.5).

Remark 3.1 It is worth to notice here that the presence of additional state constraints does not prevent the existence of solutions, provided the connectivity of the admissible state space is preserved. This is particularly important for applications in robotics. Indeed, assume that the system (3.1) models the dynamics of a robot whose space of configurations is Ω. When the robot moves amid obstacles, one represents the configurations of the robot in collision with an obstacle as closed subsets O of Ω. If the open set $\Omega \setminus O$ remains connected, the motion planning problem in that set has a solution, which means that it is possible to steer the robot from one configuration to another by avoiding the obstacles. Note that since $\Omega \setminus O$ is an open set, connectivity is equivalent to arc connectivity. The motion planning problem is then usually tackled in two steps (see [22]):

- find a curve in the free space $\Omega \setminus O$ connecting x^{initial} to x^{final} (this curve is in general not a trajectory of (3.1));
- approximate the curve by a trajectory of (3.1) close enough to be contained in the free space.

This strategy separates global topological problems from local dynamic ones. The first step is independent of the control system, it only depends on the topology of Ω and of the obstacles. It is a well modelled and understood problem of algorithmic geometry (see for instance [5, 32]). The second step may be considered as a motion planning problem without state constraints provided we use a motion planning method with a good local behaviour, namely the resulting trajectory $\gamma(\cdot; x^{\text{initial}}, u)$ converge to x^{initial} for the C^0 topology when x^{initial} converges to x^{final}. The drawback is that, imposing this property complicates the design of a motion planning method.

Strictly speaking, a solution to the motion planning problem (also called a *steering method*) is a function which associates with any pair $(x^{\text{initial}}, x^{\text{final}})$ a control u steering the system from x^{initial} to x^{final}. When the design of such an exact solution seems to be out of reach, one will rather look for an approximate solution, that is for an algorithm whose inputs are the points $(x^{\text{initial}}, x^{\text{final}})$ and a tolerance $e > 0$, and whose output is a control u steering the system from x^{initial} to a point e-close to x^{final}.

We investigate in the next section the class of nilpotent systems, a class for which exact solutions are known, and we present in the following sections approximate solutions valid for any nonholonomic system. Before these presentations let us present some ways of reducing the class of investigated systems.

Reduction of the Problem

To simplify the problem we can replace the considered nonholonomic system by another one simpler to handle, and for which the motion planning problem is equivalent. The first way to do this is to use the notion of feedback equivalence. We say that two nonholonomic systems $\dot{x} = \sum_{i=1}^{m} u_i X_i(x)$, $x \in \Omega$, and $\dot{\bar{x}} = \sum_{i=1}^{m} \bar{u}_i \bar{X}_i(\bar{x})$, $\bar{x} \in \bar{\Omega}$, are *feedback equivalent* if there exists a diffeomorphism $\Phi : \Omega \times \mathbb{R}^m \to \bar{\Omega} \times \mathbb{R}^m$ of the form,

$$\Phi(x, u) = (\phi(x), \beta(x)u), \quad \text{where } \beta(x) \text{ an invertible matrix,}$$

which transforms the first system into the second, that is

$$d\phi(x)\left(\sum_{i=1}^{m} u_i X_i(x)\right) = \sum_{i=1}^{m} \bar{u}_i \bar{X}_i(\phi(x)), \quad \text{with } \bar{u} = \beta(x)u.$$

This is equivalent to require that the sets of vector fields $\Delta = \mathrm{span}\{X_1, \ldots, X_m\}$ and $\bar{\Delta} = \mathrm{span}\{\bar{X}_1, \ldots, \bar{X}_m\}$ are diffeomorphic, or that the trajectories of both systems are diffeomorphic to each other. As a consequence of the definition, a method solving the motion planning problem for a given nonholonomic system also solves the problem for any feedback equivalent system (provided the diffeomorphism Φ is known explicitly).

A second way to simplify the problem is to desingularize the system. Assume first we want to solve the motion planning problem for (3.1) on a domain $\mathscr{V}_{\mathscr{J}} \subset \Omega$ defined by

$$\mathscr{V}_{\mathscr{J}} = \{p \in \Omega \text{ such that } \det(X_{I_1}(p), \ldots, X_{I_n}(p)) \neq 0\},$$

where $\mathscr{J} = (I_1, \ldots, I_n)$ is an n-tuple of elements of the the P. Hall basis $\mathscr{H}$. Let x^{initial} and $x^{\mathrm{final}} \in \mathscr{V}_{\mathscr{J}}$ be the initial and final points. Applying the Desingularization Algorithm presented in Sect. 2.4.2 at the point x^{final}, we obtain vector fields $\xi_1, \ldots, \xi_m$ on $\mathscr{V}_{\mathscr{J}} \times \mathbb{R}^{\tilde{n}_r - n}$ such that, first the Lie algebra $Lie(\xi_1, \ldots, \xi_m)$ is free up to step r, and second $X_i = \pi_* \xi_i$ for $i = 1, \ldots, m$, where $\pi : \mathscr{V}_{\mathscr{J}} \times \mathbb{R}^{\tilde{n}_r - n} \to \mathscr{V}_{\mathscr{J}}$ denotes the canonical projection. As a consequence, the projection by π of a trajectory $\tilde{\gamma}(\cdot; \tilde{x}^{\mathrm{initial}}, u)$ of the control system

$$\dot{\tilde{x}} = \sum_{i=1}^{m} u_i \xi_i(\tilde{x}), \quad \tilde{x} \in \mathscr{V}_{\mathscr{J}} \times \mathbb{R}^{\tilde{n}_r - n}, \tag{3.2}$$

is a trajectory of

$$\dot{x} = \sum_{i=1}^{m} u_i X_i(x), \quad x \in \mathscr{V}_{\mathscr{J}}, \tag{3.3}$$

associated with the same input, i.e., $\pi\left(\widetilde{\gamma}(\cdot\,;\widetilde{x}^{\text{initial}}, u)\right) = \gamma(\cdot\,;\pi(\widetilde{x}^{\text{initial}}), u)$. Therefore, any control u steering (3.2) from $\widetilde{x}^{\text{initial}} = (x^{\text{initial}}, 0)$ to $\widetilde{x}^{\text{final}} = (x^{\text{final}}, 0)$ also
steers (3.3) from x^{initial} to x^{final}. It is then sufficient to solve the motion planning
problem for the lifted system (3.2).

Assume now that we want to solve the motion planning problem for (3.3) on
a compact set $K \subset \Omega$ which is not included in one single domain $\mathcal{V}_{\mathscr{I}}$. Let us
denote by r the maximal value of the degree of nonholonomy on K (as mentioned in
Sect. 2.1.2, this maximum exists). For every n-tuple $\mathscr{I} = (I_1, \ldots, I_n)$ of elements
of $\mathscr{H}^r$, the domain $\mathcal{V}_{\mathscr{I}}$ is open in Ω (possibly empty) and for every $p \in \mathcal{V}_{\mathscr{I}}$, the
vectors $X_{I_1}(p), \ldots, X_{I_n}(p)$ form a basis of $\mathbb{R}^n$. Since K is compact, there exist a
finite number of n-tuples $\mathscr{I}_1, \ldots, \mathscr{I}_M$ of elements of $\mathscr{H}^r$ and connected compact
sets $K_1, \ldots, K_M$ such that $K_i \subset \mathcal{V}_{\mathscr{I}_i}$ for $i = 1, \ldots, M$, and

$$K \subset \bigcup_{i=1}^{M} K_i. \tag{3.4}$$

It is clearly sufficient to solve the motion planning on each of these compact sets
$K_i \subset \mathcal{V}_{\mathscr{I}_i}$, where we can apply the method by lifting described above.

As a conclusion, we can reduce the motion planning problem to systems (3.1)
such that the vectors fields $X_1, \ldots, X_m$ generate a free up to step r Lie algebra. If
moreover (3.1) is nilpotent, by Remark 2.9 we can also assume that $(X_1, \ldots, X_m)$
is given in the canonical form.

3.2 Nilpotent Systems

A *nilpotent system* is a nonholonomic systems (3.1) such that the vector fields
$X_1, \ldots, X_m$ generate a nilpotent Lie algebra. A system (3.1) is *nilpotentizable* if
it is feedback equivalent to a nilpotent system.

Nilpotent systems play a very important role for motion planning purposes. First,
nilpotentizable systems form the widest class of nonholonomic system for which an
exact solution to the motion planning problem is known. Second, we have seen in
Sect. 2.1.3 that every nonholonomic system admits first-order approximations which
are nilpotent systems. Thus when using Newton like methods to solve the problem,
we have to deal with these nilpotent systems.

Note however that most of the nonholonomic systems are not nilpotent, and not
nilpotentizable either. In fact, the class of nilpotentizable systems is clearly a non
generic class of nonholonomic systems when the dimension of the state space is
large enough. In addition, there are no criteria for deciding whether a system is
nilpotentizable.

We give in this section a solution to the motion planning problem for nilpotent
systems which uses sinusoidal controls. We present first the method in the simple
case of chained systems in Sect. 3.2.1, and then in the general case in Sect. 3.2.2. We
finally discuss other methods in Sect. 3.2.3.

3.2.1 The Case of Chained Systems

In order to better understand the use of sinusoidal controls, we consider first the particular case of a *chained system*, which is a nonholonomic system on $\mathbb{R}^n$ in the following form,

$$
\begin{aligned}
\dot{x}_1 &= u_1 \\
\dot{x}_2 &= u_2 \\
\dot{x}_3 &= x_2 u_1 \\
\dot{x}_4 &= x_3 u_1 \\
&\ \ \vdots \\
\dot{x}_n &= x_{n-1} u_1.
\end{aligned}
\tag{3.5}
$$

Equivalently, $\dot{x} = u_1 X_1(x) + u_2 X_2(x)$, where the vector fields X_1 and X_2 are given by

$$
X_1 = \partial_{x_1} + \sum_{i=3}^{n} x_{i-1}\partial_{x_i}, \qquad X_2 = \partial_{x_2}.
$$

Such a system is clearly nilpotent, the only nonzero brackets of X_1 and X_2 being

$$
(\operatorname{ad}X_1)^i X_2 = (-1)^i \partial_{x_{i+2}}, \qquad i = 1, \ldots, n - 2,
$$

where we write $(\operatorname{ad}X_1)X_2$ for $[X_1, X_2]$, $(\operatorname{ad}X_1)^2 X_2$ for $(\operatorname{ad}X_1)((\operatorname{ad}X_1)X_2)$, etc. It also satisfies the Chow Condition.

The particular structure of (3.5) suggests to control the system one component after the other. The crucial point of this strategy is to ensure that if a component x_k is moved during a period $[0, T]$, then none of the components x_i with $i < k$ is moved, that is $x_i(0) = x_i(T)$. The use of sinusoidal controls with integer frequencies is well suited to such a strategy because of the usual properties:

$$
\int_0^{2\pi} \sin \omega t \, dt = 0 \ \text{ for all } \omega \in \mathbb{Z},
\tag{3.6}
$$

$$
\int_0^{2\pi} \cos \omega t \, dt =
\begin{cases}
0 & \text{if } \omega \in \mathbb{Z} \text{ and } \omega \neq 0, \\
1 & \text{if } \omega = 0.
\end{cases}
\tag{3.7}
$$

Consider then a control $u(t) = (u_1(t), u_2(t))$, $t \in [0, 2\pi]$, of the form

$$
u_1(t) = a \sin \omega_1 t, \qquad u_2(t) = b \cos \omega_2 t.
$$

Due to the particular structure of (3.5), for every $i \geq 3$ the dynamics $\dot{x}_i$ is a linear combination of sinusoids whose frequencies are

$$\omega_2 + N\,\omega_1, \qquad \text{where } N \in \mathbb{Z}, \quad |N| \leq i - 2. \tag{3.8}$$

Moreover the sinusoid with frequency $\omega_2 + (i - 2)\,\omega_1$ appears with a nonzero coefficient (if a and b are nonzero).

Assume that we want to move x_k to its desired value without changing the values of x_i, for every $i < k$. The relations (3.6) and (3.7) imply that we must choose ω_1 and ω_2 so that one cosine term with null frequency (i.e. a constant term) appears in the dynamic component $\dot{x}_k$. In addition, no component x_i with $i < k$ should verify this condition in order to ensure $x_i(0) = x_i(2\pi)$. For instance, a simple choice guaranteeing both properties is $\omega_1 = 1$ et $\omega_2 = k - 2$.

The argument above translates to a complete algorithm (Algorithm 3.1) which gives an exact solution to the motion planning problem for the system (3.5).

Algorithm 3.1 Steering method for chained systems

1: move x_1 and x_2 to their final positions (in straight line for instance);

2: for every $k \geq 1$, move x_{k+2} from its current value x_{k+2}^k to its target value x_{k+2}^{final} by using

$$u(t) = (a \sin t, b \cos kt), \quad t \in [0, 2\pi],$$

where the parameters a and b verify

$$x_{k+2}^{\text{final}} - x_{k+2}^k = \frac{(a/2)^k b}{k!} 2\pi.$$

3.2.2 Sinusoidal Controls

Consider a nilpotent system on $\mathbb{R}^n$ defined by vector fields $(X_1, \ldots, X_m)$. As noticed in Sect. 3.1, up to a desingularization we can reduce the problem to a system satisfying the following three assumptions (all notions have been introduced in Sect. 2.4):

- the Lie algebra $Lie(X_1, \ldots, X_m)$ is free up to step r, where r is its nilpotency step; as a consequence, $n = \tilde{n}_r$;
- $(X_1, \ldots, X_m)$ is given in the canonical form in some coordinates x, i.e. the control system is written as follows,

$$\begin{aligned}
\dot{x}_i &= u_i, & \text{if } i = 1, \ldots, m; \\
\dot{x}_I &= \tfrac{1}{k!} x_{I_{j_1}} \dot{x}_{I_{j_2}}, & \text{if } I = \mathrm{ad}_{I_{j_1}}^k I_{j_2}, \quad I_{j_1}, I_{j_2} \in \mathcal{H}^r,
\end{aligned} \tag{3.9}$$

where the components of x are numbered by the elements of $\mathcal{H}^r$;

- the goal point x^{final} in the motion planning problem is always the origin 0 of $\mathbb{R}^{\tilde{n}_r}$ in coordinates x.

As we did for chained systems, we will choose control functions in the form of linear combinations of sinusoids with integer frequencies.

We first note that if every component u_i of the control u in (3.9) is a linear combination of sinusoids with integer frequencies, then every component x_I of a solution of (3.9) is also a linear combination of sinusoids with integer frequencies, the latter being linear combinations of the frequencies involved in u. One may therefore expect to move some components during a 2π time-period without modifying others if the frequencies in u are properly chosen.

Due to the triangular form of (3.9), it is reasonable to expect to move the components of x one after another according to the order $\prec$ induced by the P. Hall basis. In that case, one must ensure that, after each 2π-period of control process, the component under consideration arrives to its preassigned position, and all "smaller" (w.r.t. $\prec$) components return to their initial values. However, it appears that all the components cannot be moved independently by using sinusoids. To take this into account, we introduce the following notion of equivalence.

Definition 3.2 For $I \in \mathscr{L}(1, \ldots, m)$ and $i \in \{1, \ldots, m\}$, we denote by $|I|_i$ the number of times i occurs in I. Two components x_I and x_J, $I, J \in \mathscr{H}^r$, are said to be *equivalent* if $|I|_i = |J|_i$ for $i = 1, \ldots, m$. The equivalent classes of components will be denoted by

$$\mathscr{E}(\ell_1, \ldots, \ell_m) = \{x_I \ : \ |I|_i = \ell_i, \text{ for } i = 1, \ldots, m\}.$$

We will see below that the frequencies occurring in the dynamics of x_I only depend on the equivalence class of x_I, and not on the structure of the bracket $I \in \mathscr{H}^r$. Therefore, the equivalent components cannot be moved separately by using sinusoids.

The equivalences classes inherit an order from the one of the P. Hall basis. A class $\mathscr{E}(\ell_1, \ldots, \ell_m)$ is said to be smaller than $\mathscr{E}(\tilde{\ell}_1, \ldots, \tilde{\ell}_m)$ if the smallest element (w.r.t. $\prec$) in $\mathscr{E}(\ell_1, \ldots, \ell_m)$ is smaller than the smallest one in $\mathscr{E}(\tilde{\ell}_1, \ldots, \tilde{\ell}_m)$, and we write (by abuse of notation) $\mathscr{E}(\ell_1, \ldots, \ell_m) \prec \mathscr{E}(\tilde{\ell}_1, \ldots, \tilde{\ell}_m)$.

The components of x may be partitioned in equivalent classes $\mathscr{E}^1, \mathscr{E}^2, \ldots, \mathscr{E}^{\tilde{N}}$, sorted by increasing order. Our control strategy consists in displacing these equivalence classes one after another according to the order $\prec$ by using sinusoidal inputs. For every $i = 1, \ldots, \tilde{N}$, the key point is to determine how to construct an input u^i defined on $[0, 2\pi]$ such that the two following conditions are verified:

(C1) under the action of u^i, every element of $\mathscr{E}^i$ reaches its preassigned value at $t = 2\pi$;
(C2) under the action of u^i, for all $j < i$, every element of $\mathscr{E}^j$ returns at $t = 2\pi$ to its value taken at $t = 0$.

Once one knows how to construct an input u^i verifying (C1) and (C2) for every $i = 1, \ldots, \tilde{N}$, it suffices to take the concatenation of these inputs in order to steer the complete system to the goal. The resulting control is defined on the interval $[0, 2\widetilde{N}\pi]$ by

$$u^1 * \cdots * u^{\widetilde{N}}(t) = u^i(t - 2(i-1)\pi), \tag{3.10}$$

for $t \in [2(i-1)\pi, 2i\pi]$ and $i \in \{1, \ldots, \widetilde{N}\}$.

Choice of Frequencies for a Class $\mathscr{E}^i$

Let us fix an equivalence class $\mathscr{E}^i$. According to the strategy described just above, we have to choose the frequencies appearing in the sinusoidal control u^i in such a way that Conditions (C1) and (C2) are verified. For sake of clarity, we only treat the case $m = 2$, the reader is referred to [9] on how to adapt the method to greater values of m.

Assume first that $\mathscr{E}^i$ contains only one component, denoted by x_I. Set $m_1 = |I|_1$, and $m_2 = |I|_2$, so that $\mathscr{E}^i = \mathscr{E}(m_1, m_2)$. We look for the control u^i under the form,

$$u^i_1(t) = \cos \omega_1 t, \quad u^i_2(t) = \cos \omega_2 t + a \cos(\omega_2^* t - \varepsilon \frac{\pi}{2}), \quad t \in [0, 2\pi], \tag{3.11}$$

where $\omega_1, \omega_2, \omega_2^*$ are positive integers, $\varepsilon \in \{0, 1\}$, and a is a real number.

By induction, one obtains that, for every $J \in \mathscr{H}^r$ such that $J \preceq I$, the dynamics $\dot{x}_J$ is a linear combination of cosine functions,

$$\cos \left(\left(\ell_1 \omega_1 + \ell_2 \omega_2 + \ell_3 \omega_2^* \right) t - (\ell_3 \varepsilon + \ell_1 + \ell_2 + \ell_3 - 1) \frac{\pi}{2} \right), \tag{3.12}$$

where $\ell_1, \ell_2, \ell_3 \in \mathbb{Z}$ satisfy $|\ell_1| \leq m_1$, $|\ell_2| + |\ell_3| \leq m_2$. In particular, the term

$$\cos \left(\left(m_1 \omega_1 + (m_2 - 1)\omega_2 - \omega_2^* \right) t - (-\varepsilon + m_1 + m_2 - 1) \frac{\pi}{2} \right)$$

occurs in $\dot{x}_I$ with a nonzero coefficient that depends linearly on a.

Now, note that integrating between 0 and 2π a function of the form $\cos(\omega t + \varepsilon \frac{\pi}{2})$ with $\omega \in \mathbb{Z}$ and $\varepsilon \in \mathbb{N}$ always gives 0 except if $\omega = 0$ and $\varepsilon = 0 \mod 2$. Therefore, in order to obtain a non trivial contribution in the component x_I, its derivative $\dot{x}_I$ must contain some cosine functions of the form (3.12) verifying the following condition

$$\begin{cases} \ell_1 \omega_1 + \ell_2 \omega_2 + \ell_3 \omega_2^* = 0, \\ \ell_3 \varepsilon + \ell_1 + m_2 + \ell_3 - 1 \equiv 0 \mod 2. \end{cases} \tag{3.13}$$

Moreover this condition shall not be satisfied by any cosine function appearing in $\dot{x}_J$, $J \prec I$, so as to prevent non trivial contributions in the component x_J. A way to ensure both properties is to impose

$$\begin{cases} \omega_2^* = m_1\omega_1 + (m_2 - 1)\omega_2, \\ \varepsilon = m_1 + m_2 - 1 \mod 2, \end{cases} \tag{3.14}$$

and

$$\omega_2 > (m_1 + m_2)m_1\omega_1. \tag{3.15}$$

In that case,

- Condition (3.13) is never satisfied by a cosine function appearing in $\dot{x}_J$, $J \prec I$;
- among the cosine function of the form (3.12) appearing in $\dot{x}_I$, the function with $\ell_1 = m_1$, $\ell_2 = m_2 - 1$, and $\ell_3 = -1$ is the only one that verifies (3.13);
- $x_I(2\pi) - x_I(0)$ is proportional to a with a ratio $f_I = f_I(\omega_1, \omega_2)$.

Let us precise the last point. For every element $J \preceq I$, we set $m_1^J = |J|_1, m_2^J = |J|_2$, and, in the decomposition of $\dot{x}_J$ as a linear combination of cosine functions, we denote by F_J the coefficient in front of the term

$$\cos\left((m_1^J\omega_1 + (m_2^J - 1)\omega_2 - \omega_2^*)t - (m_1^J + m_2^J - 1 - \varepsilon)\frac{\pi}{2}\right),$$

and by g_J the one in front of the term

$$\cos\left((m_1^J\omega_1 + m_2^J\omega_2)t - (m_1^J + m_2^J - 1)\frac{\pi}{2}\right).$$

Then F_J is proportional to a and we write $F_J = af_J(\omega_1, \omega_2)$. Moreover $g_J = g_J(\omega_1, \omega_2)$ is a never vanishing function of (ω_1, ω_2) and the quotient $\alpha_J = f_J/g_J$ verifies the following inductive formula:

- $\alpha_1 = 0, \alpha_2 = 1$;
- if $J = [J_1, J_2]$, then

$$\alpha_J = \frac{m_1^{J_1}\omega_1 + m_2^{J_1}\omega_2}{m_1^{J_1}\omega_1 + (m_2^{J_1} - 1)\omega_2 - \omega_2^*}\alpha_{J_1} + \alpha_{J_2}.$$

This formula implies that α_I is a rational function of (ω_1, ω_2) which can be proved to be non identically zero. Hence α_I, and so f_I, is nonzero on $\mathbb{R}^2$ minus a finite number of algebraic hypersurfaces. In other words, for almost all ω_1, ω_2 satisfying (3.15), $x_I(2\pi) - x_I(0)$ is proportional to a with a nonzero ratio.

As a consequence, by choosing properly ω_1, ω_2 and a, the control u^i steers, during $[0, 2\pi]$, the component x_I from any initial value to any preassigned final value without modifying any component x_J with $J \prec I$.

Assume now that the equivalence class $\mathscr{E}^i = \mathscr{E}(m_1, m_2)$ contains two components x_{I_1} and x_{I_2}. This situation first occurs for Lie brackets of length 5, for instance in $\mathscr{E}(2, 3)$ with $I_1 = [2, [1, [1, [1, 2]]]]$ and $I_2 = [[1, 2], [1, [1, 2]]]$. If one chooses frequencies verifying the resonance condition (3.14) for $\dot{x}_{I_1}$, the same resonance occurs in $\dot{x}_{I_2}$. Such two components cannot be independently steered by using resonance. The idea is then to move simultaneously these components. For instance, one can choose u^i as follows,

$$
\begin{aligned}
u_1^i(t) &= \cos \omega_1 t, \\
u_2^i(t) &= \cos \omega_2 t + a_I \cos \omega_2^* t + \cos \omega_3 t + a_J \cos \omega_3^* t,
\end{aligned}
$$

with $\omega_1 = 1$, ω_2 satisfying (3.15),

$$
\omega_2^* = (m_2 - 1)\omega_2 + m_1 \omega_1, \quad \omega_3^* = (m_2 - 1)\omega_3 + m_1 \omega_1,
$$

and ω_3 large enough to guarantee Condition (C2). After explicit integration of (3.9), one obtains

$$
\begin{pmatrix} x_{I_1}(2\pi) - x_{I_1}(0) \\ x_{I_2}(2\pi) - x_{I_2}(0) \end{pmatrix} = \begin{pmatrix} f_{I_1}(\omega_1, \omega_2) & f_{I_1}(\omega_1, \omega_3) \\ f_{I_2}(\omega_1, \omega_2) & f_{I_2}(\omega_1, \omega_3) \end{pmatrix} \begin{pmatrix} a_{I_1} \\ a_{I_2} \end{pmatrix} = A \begin{pmatrix} a_{I_1} \\ a_{I_2} \end{pmatrix},
$$

where f_{I_1} and f_{I_2} are two rational functions. Thus, u^i controls exactly and simultaneously x_{I_1} and x_{I_2}, provided that the matrix A is invertible.

Let us generalize this strategy. Let $x_{I_1}, \ldots, x_{I_N}$ be the elements of the equivalence class $\mathcal{E}^i = \mathcal{E}(m_1, m_2)$. We look for a control u^i of the form

$$
\begin{cases}
u_1^i = \displaystyle\sum_{k=1}^{N} \sum_{j=1}^{m_1} \cos \omega_{1k}^j t, \\[2ex]
u_2^i = \displaystyle\sum_{k=1}^{N} \left(\sum_{j=1}^{m_2-1} \cos \omega_{2k}^j t + a_k \cos(\omega_{2k}^* t - \varepsilon \frac{\pi}{2}) \right),
\end{cases} \qquad t \in [0, 2\pi], \qquad (3.16)
$$

where all the frequencies ω_{lk}^j, ω_{lk}^* are positive integers, ε equals 0 or 1, and $a_1, \ldots, a_N$ are real numbers. We impose "resonance relations",

$$
\begin{cases}
\omega_{2k}^* = \displaystyle\sum_{j=1}^{m_1} \omega_{1k}^j + \sum_{j=1}^{m_2-1} \omega_{2k}^j, & \text{for } k = 1, \ldots, N, \\[2ex]
\varepsilon = m_1 + m_2 - 1 \mod 2,
\end{cases} \qquad (3.17)
$$

and the following inequalities, for $k = 1, \ldots, N - 1$,

$$
\begin{cases}
\omega_{1k}^{j+1} > m_1 \omega_{1k}^j, & j = 1, \ldots, m_1, \\
\omega_{2k}^1 > m_1 \omega_{1k}^{m_1}, & \\
\omega_{2k}^j > m_2 \omega_{2k}^{j-1} + m_1 \omega_{1k}^{m_1}, & j = 2, \ldots, m_2 - 1, \\
\omega_{1k+1}^1 > m_2 \omega_{2k}^{m_2-1} + m_1 \omega_{1k}^{m_1}.
\end{cases} \qquad (3.18)
$$

We then obtain that, if $x_J \in \mathcal{E}^j$ with $j < i$, then $x_J(2\pi) - x_J(0) = 0$. Moreover,

$$
\begin{pmatrix} x_{I_1}(2\pi) - x_{I_1}(0) \\ \vdots \\ x_{I_N}(2\pi) - x_{I_N}(0) \end{pmatrix} = A(\omega_{11}^1, \ldots, \omega_{2N}^{m_2-1}) \begin{pmatrix} a_1 \\ \vdots \\ a_N \end{pmatrix}
\tag{3.19}
$$

where $A(\omega_{11}^1, \ldots, \omega_{2N}^{m_2-1})$ is a $(N \times N)$ matrix whose coefficients are rational functions of the frequencies. One can then show that, for almost all[1] frequencies ω_{lk}^j satisfying (3.18), the matrix $A(\omega_{11}^1, \ldots, \omega_{2N}^{m_2-1})$ is invertible (see [9] for the proof).

As a consequence, we can choose $a_1, \ldots, a_N$ so that the corresponding control u^i steers, during $[0, 2\pi]$, the components $x_{I_1}, \ldots, x_{I_N}$ of $\mathcal{E}^i$ from any initial values to any preassigned final values without modifying any component x_J in $\mathcal{E}^j$ with $j < i$.

Let us summarize the discussion above. We have seen that it is possible to construct a control u^i in the form (3.16) that satisfy Conditions (C1) and (C2). Our construction involves $N(m_1 + m_2)$ integer frequencies that are widely spaced (condition (3.18)) in order to prevent contributions in components x_I belonging to "smaller" equivalence classes, and that verify a resonance condition (equation (3.17)) which implies the linear relation (3.19).

However this construction tends to produce high frequencies while it is desirable to find smaller ones for practical use. Therefore, the solution we recommend is to implement an algorithm that will search for the frequencies appearing in the expression (3.16) of the control u^i. This algorithm would test iteratively all $N(m_1 + m_2)$-tuple of integer satisfying the resonance condition (3.17), and would search for the smallest ones which prevent contributions in components x_I belonging to classes $\mathcal{E}^j$ with $j < i$, and which produce an invertible matrix A. The discussion above guarantees the finiteness of such an algorithm.

Note also that we could test in that algorithm controls u^i involving less frequencies. We have seen that 3 frequencies are enough when $N = 1$ whatever the values of m_1, m_2. We conjecture that, for $N \geq 1$, $3N$ frequencies suffice, instead of $N(m_1 + m_2)$, and more generally $N(m + 1)$ when m is greater than 2.

Anyway, for each equivalence class $\mathcal{E}^i = \{I_1, \ldots, I_N\}$, the result of these off-line computations will be a control law $u^i[a] : [0, 2\pi] \to \mathbb{R}^m$ depending on a parameter $a \in \mathbb{R}^N$, and an invertible matrix A_i such that:

- $u^i[a]$ is a linear combination of cosine functions with integer frequencies ($u^i[a]$ is defined by (3.16) in the above construction);
- the trajectory associated with $u^i[a]$ satisfy $x_J(2\pi) = x_J(0)$ if $J \in \mathcal{E}^j$ with $j < i$, and

---

[1] More precisely, for all points with integer coordinates in $\mathbb{R}^{N(m_1+m_2)}$ minus a finite number of algebraic hypersurfaces, that number being a function of N, m_1 and m_2.

$$\begin{pmatrix} x_{I_1}(2\pi) - x_{I_1}(0) \\ \vdots \\ x_{I_N}(2\pi) - x_{I_N}(0) \end{pmatrix} = A_i a.$$

Design of the Steering Law

We are now in a position to construct a steering method for the nilpotent system defined by $X_1, \ldots, X_m$. Note that since $X_1, \ldots, X_m$ are assumed to be given in canonical form, the method only depends on the number of controlled vector fields m and on the degree of nonholonomy r. The method is devised so that the produced trajectories satisfy the topological property of Remark 3.1, namely the resulting trajectory $\gamma(\cdot; x^{\text{initial}}, u)$ converge to x^{initial} for the C^0 topology when x^{initial} converges to x^{final}. Note that we have assumed that x^{final} is always the origin.

Let us fix first some notations. Remind that the components of $x \in \mathbb{R}^{\tilde{n}_r}$ are partitioned in equivalent classes $\mathscr{E}^1, \mathscr{E}^2, \ldots, \mathscr{E}^{\tilde{N}}$, sorted by increasing order. With every equivalence class $\mathscr{E}^i$ is associated a parameterized control law $u^i[a]$ and an invertible matrix A_i. We set $B_i = A_i^{-1}$ and $N_i = \text{Card}(\mathscr{E}^i)$, $i = 1, \ldots, \tilde{N}$. For $x \in \mathbb{R}^{\tilde{n}_r}$, we will use $[x]_{j,\ldots,k}$ with $1 \le j < k \le \tilde{n}_r$ to denote the vector $(x_j, \ldots, x_k)$, and $\|x\|_0$ to denote the pseudo-norm of x defined by the free weights. We also use $\delta_\lambda(x) = (\lambda^{w_1} x_1, \ldots, \lambda^{w_n} x_n)$ to denote the weighted dilation with parameter λ.

Algorithm 3.2 produces a control $\text{Steer}_{m,r}(x^{\text{initial}})$ that steers the canonical form from an initial point $x^{\text{initial}} \in \mathbb{R}^{\tilde{n}_r}$ to the origin.

Algorithm 3.2 Construction of the control law $\text{Steer}_{m,r}(x^{\text{initial}})$

Require: $B_1, \ldots, B_{\tilde{N}}$, and $N_1, \ldots, N_{\tilde{N}}$;

1: $\lambda := \|x^{\text{initial}}\|_0$;

2: $x^{\text{new}} := \delta_{\frac{1}{\lambda}}(x^{\text{initial}})$;

3: $u_{\text{norm}} := 0$;

4: $j := 0$;

5: **for** $i = 1, \ldots, \tilde{N}$ **do**

6: $\quad x := [x^{\text{new}}]_{j+1,\ldots,j+N_i}$;

7: $\quad a^i := B_i x$;

8: $\quad$ construct $u^i = u^i[a^i]$ by Formula (3.16);

9: $\quad x^{\text{new}} := \gamma(2\pi; x^{\text{new}}, u^i)$;

10: $\quad u_{\text{norm}} := u_{\text{norm}} * u^i$;

11: $\quad j = j + N_i$;

12: **return** $u := \lambda \hat{u}_{\text{norm}}$.

Remark 3.2 Once the frequencies and matrices are obtained by an off-line computation, the on-line computation $\text{Steer}_{m,r}$ contains only a series of matrix multiplications, and the computations of $\gamma(2\pi; x^{\text{new}}, u^i)$. The latter do not require to solve differential equations but are obtained simply by successive integrations of u^i, because the system is in triangular form. Thus all computations in $\text{Steer}_{m,r}$ can be performed quickly on-line without any numerical difficulty.

Proposition 3.1 *For every* $x^{\text{initial}} \in \mathbb{R}^{\tilde{n}_r}$, *the control* $\text{Steer}_{m,r}(x^{\text{initial}})$ *steers System* (3.9) *from* x^{initial} *to* $x^{\text{final}} = 0$. *Moreover, there exists a constant* $C > 0$ *such that*

$$\|\text{Steer}_{m,r}(x^{\text{initial}})\|_{L^1} \le C d_0(0, x^{\text{initial}}), \qquad \forall\, x^{\text{initial}} \in \mathbb{R}^{\tilde{n}_r}, \tag{3.20}$$

where we use d_0 *to denote the sub-Riemannian distance on* $\mathbb{R}^{\tilde{n}_r}$ *defined by* $X_1, \ldots, X_m$.

Note that the length of the trajectory $\gamma(\cdot; x^{\text{initial}}, u)$ associated with the control $u = \text{Steer}_{m,r}(x^{\text{initial}})$ is

$$\text{length}\left(\gamma(\cdot; x^{\text{initial}}, u)\right) = \|\text{Steer}_{m,r}(x^{\text{initial}})\|_{L^1}.$$

Hence (3.20) implies that this length tends to 0 as $x^{\text{initial}} \to 0$, which is the topological property required in Remark 3.1. We will see in Sect. 3.3 that this property is also related to the notion of *sub-optimal laws.*

Proof The fact that the procedure described by the Lines 5–12 in Algorithm 3.2 produces an input $\hat{u}_{\text{norm}}$ steering System (3.9) from $\delta_{\frac{1}{\lambda}}(x^{\text{initial}})$ to 0 results from the choice of frequencies discussed previously. We also note that, due to the homogeneity of System (3.9), if an input u steers this system from x to 0, then, for every $\lambda > 0$, the input λu steers it from $\delta_\lambda(x)$ to 0. Therefore, the input $\text{Steer}_{m,r}(x^{\text{initial}})$ steers (3.9) from x^{initial} to 0.

Let us now show (3.20). Let $x \in \mathbb{R}^{\tilde{n}_r}$ and set $\lambda = \|x\|_0$. Then $x = \delta_\lambda(x_{\text{norm}})$ where x_{norm} belongs to the sphere $S(0, 1) = \{y : \|y\|_0 = 1\}$. Setting $u(x) = \text{Steer}_{m,r}(x)$ we have

$$\|u(x)\|_{L^1} = \|\lambda u(x_{\text{norm}})\|_{L^1} = \lambda \|u(x_{\text{norm}})\|_{L^1} \le \lambda \sup_{y \in S(0,1)} \|u(y)\|_{L^1}.$$

Note that $d_0(0, \cdot)$ and $\|\cdot\|_0$ at 0 are both homogeneous of degree 1 with respect to the dilation $\delta_t(\cdot)$, so there exists a constant $\tilde{C} > 0$ such that $\lambda \le \tilde{C} d(0, x)$. Moreover $\sup_{y \in S(0,1)} \|u(y)\|_{L^1}$ is bounded since $y \mapsto u(y)$ is continuous and $S(0, 1)$ compact. Inequality (3.20) follows. $\qquad\qquad\square$

Remark 3.3 The control we constructed are C^∞ during each time interval $[2i\pi, 2(i+1)\pi]$, for $i = 1, \ldots, \tilde{N} - 1$, but not globally continuous on $[0, 2\tilde{N}\pi]$ due to discontinuity at $t = 2\pi, 4\pi, \ldots, 2(\tilde{N} - 1)\pi$. It is however not difficult to devise continuous controls by using interpolation techniques. We illustrate the idea with a simple example. Assume that we use u^i and u^j defined by

$$u_1^i(t) = \cos \omega_{1i} t,$$

$$u_2^i(t) = \cos \omega_{2i} t + a^i \cos(\omega_{2i}^* t + \varepsilon^i \frac{\pi}{2}), \qquad t \in [2(i-1)\pi, 2i\pi],$$

$$u_1^j(t) = \cos \omega_{1j} t,$$

$$u_2^j(t) = \cos \omega_{2j} t + a^j \cos(\omega_{2j}^* t + \varepsilon^j \frac{\pi}{2}), \qquad t \in [2(j-1)\pi, 2j\pi],$$

to steer two consecutive classes $\mathscr{E}^i$ and $\mathscr{E}^j$ (i.e. $j = i + 1$) which are both of cardinal equal to 1. The concatenation $u^i * u^j$ is not continuous, so we will construct a modification $\widetilde{u}^j$ of u^j in such a way that $u^i * \widetilde{u}^j$ is continuous, i.e.

$$u_1^i(2\pi) = \widetilde{u}_1^j(2\pi), \tag{3.21}$$

$$u_2^i(2\pi) = \widetilde{u}_2^j(2\pi). \tag{3.22}$$

To ensure (3.21), we take

$$\widetilde{u}_1^j(t) = u_1^i(2\pi) \cos \omega_{1j} t. \tag{3.23}$$

For (3.22), we distinguish two cases:

- if $\varepsilon^j = 1$, we can take

$$\widetilde{u}_2^j(t) = u_2^i(2\pi) \cos \omega_{2j} t + a^j \cos(\omega_{2j}^* t - \frac{\pi}{2}); \tag{3.24}$$

- if $\varepsilon^j = 0$, we add to u_2^j a frequency ω_c which is large enough to avoid any additional resonances,

$$\widetilde{u}_2^j(t) = \cos \omega_{2j} t + a^j \cos \omega_{2j}^* t + (u_2^i(2\pi) - a^j - 1) \cos \omega_c t. \tag{3.25}$$

By construction, the new input $u^i * \widetilde{u}^j$ is continuous over the time interval $[2i\pi, 2j\pi]$, and steers the components in $\mathscr{E}^i$ and $\mathscr{E}^j$ to the same values as $u^i * u^j$ does.

It is clear that this idea of interpolation by adding suitable frequencies can be used to construct continuous inputs over the entire control period $[0, 2\widetilde{N}\pi]$. In fact, by using more refined interpolations, one can get inputs of class C^k for any arbitrary finite integer k.

3.2.3 Other Methods for Nilpotent Systems

Besides sinusoids, other classes of functions may be used to give exact solutions to the motion planning problem for nilpotent systems, in particular polynomials and piecewise constants functions.

Polynomial Inputs

Assume that the nilpotent system $\dot{x} = \sum_{i=1}^{m} u_i X_i(x)$ is polynomial and in triangular form, that is,

$$\dot{x}_j = \sum_i u_i f_{ij}(x_1, \ldots, x_{j-1}), \quad j = 1, \ldots, n,$$

where every function f_{ij} is polynomial. Chained systems and systems in canonical form satisfy this assumption, as well as nilpotent approximations expressed in privileged coordinates (see (2.6) in Sect. 2.1.3). Actually, any analytic nilpotent system can be put locally in this form [17].

This structure allows to compute easily the trajectories: given a control function $u(t)$, the coordinates $x_j(t)$ may be computed by integration one after the other. Let us choose the controls as parameterized polynomial functions, for instance,

$$u_i(t) = \sum_{k=0}^{N} a_{ik} t^k, \quad i = 1, \ldots, m, \tag{3.26}$$

with a parameter $a = (a_{10}, \ldots, a_{m0}, \ldots, a_{1N}, \ldots, a_{mN}) \in \mathbb{R}^{(N+1)m}$, N being a large enough integer. With such a control every coordinate $x_j(t)$ is a polynomial function of t, of the parameter a, and of the initial values $x_1(0), \ldots, x_j(0)$. In particular there holds,

$$x_j(1) = P_j(a, x(0)), \quad \text{where } P_j \text{ is a polynomial function.}$$

Thus, in order to obtain a control of the form (3.26) steering the system from x^{initial} to x^{final}, it is necessary to solve the algebraic system,

$$P_j(a, x^{\text{initial}}) - x_j^{\text{final}} = 0, \quad j = 1, \ldots, n,$$

where the unknown is the parameter $a \in \mathbb{R}^{(N+1)m}$ (of course one must choose N such that $(N + 1)m \geq n$). Unfortunately the size and the degree of this algebraic system increase exponentially with respect to the dimension n and to the degree of nilpotency r, and there does not exist a general efficient method to solve it. Even the existence of solutions is a non trivial issue. However this method may be useful when n or r is rather small (seemingly when $n \leq 5$ or $r = 2$).

Piecewise Constant Controls

Consider a nilpotent system of step r on $\mathbb{R}^n$ defined by analytic vector fields $(X_1, \ldots, X_m)$. We fix an initial point $x^{\text{initial}} \in \mathbb{R}^n$ and an open neighbourhood U of this point. We make the following assumption.

Assumption (A) The system is provided with:

(i) a family of brackets $X_{I_1}, \ldots, X_{I_{\widetilde{n}}}, \widetilde{n} \geq n$, whose values at x^{initial} generates the linear space $Lie(X_1, \ldots, X_m)(x^{\text{initial}})$;
(ii) a mapping $x \in U \mapsto q = q(x) \in \mathbb{R}^{\widetilde{n}}$ such that

$$x = \exp(q_{\widetilde{n}} X_{I_{\widetilde{n}}}) \circ \cdots \circ \exp(q_1 X_{I_1})(x^{\text{initial}}). \tag{3.27}$$

Let x^{final} be the goal point and $q = q(x^{\text{final}})$. Given a control function u, we denote by $S^u(t)$ the flow of the time-dependant vector field $\sum_i u_i X_i$. The principle of the method is to produce separately and sequentially, for $i = 1, \ldots, \widetilde{n}$, control functions u^i defined on $[0, T]$ such that $S^{u^i}(T)$ equals $\exp(q_i X_{I_i})$ up to flows of brackets of bigger length, the latter modifying the remaining factors $\exp(q_j X_{I_j})$ with $j > i$ (the I_j are ordered by increasing length). This is not difficult to achieve with piecewise constant controls thanks to the Campbell-Hausdorff formula (see Sect. A.1). We illustrate this idea with an example.

Assume $m = 2$, $I_1 = 1$, $I_2 = 2$ and $I_3 = [1, 2]$. For simplicity we also suppose $q_3 > 0$. Define the controls $u^1(t) = (q_1, 0)$ and $u^2(t) = (0, q_2), t \in [0, 1]$. The associated flows at time $t = 1$ satisfy $S^{u^1}(1) = \exp(q_1 X_{I_1})$ and $S^{u^2}(1) = \exp(q_2 X_{I_2})$. Define the piecewise control u^3 as,

$$u^3(t) = \begin{cases} (\sqrt{q_3}, 0) & \text{for } t \in [0, 1], \\ (0, \sqrt{q_3}) & \text{for } t \in [1, 2], \\ (-\sqrt{q_3}, 0) & \text{for } t \in [2, 3], \\ (0, -\sqrt{q_3}) & \text{for } t \in [3, 4]. \end{cases}$$

The flow at time $t = 4$ $S^{u^3}(4)$ equals $\phi_{q_3}^{[1,2]} = \exp(R) \circ \exp(q_3[X_1, X_2])$, where R is a Lie polynomial that involves Lie brackets of length greater than 2 (see Remark A.1).

If the degree of nilpotency r is equal to 2 (and so $n = 3$), then $R = 0$ and the concatenation of the controls $u^1 * u^2 * u^3$ steers the system from x^{initial} to

$$\exp(q_3 X_{I_3}) \circ \exp(q_2 X_{I_2}) \circ \exp(q_1 X_{I_1})(x^{\text{initial}}) = x^{\text{final}}.$$

If $r > 2$, we can rewrite the product $\exp(q_{\tilde{n}} X_{I_{\tilde{n}}}) \circ \cdots \circ \exp(q_4 X_{I_4}) \circ \exp(R)$ as $\exp(\overline{q}_{\tilde{n}} X_{I_{\tilde{n}}}) \circ \cdots \circ \exp(\overline{q}_4 X_{I_4})$, and we can iterate the construction to generate the term $\exp(\overline{q}_4 X_{I_4})$. This procedure finishes in a finite number of steps and produces an exact solution to the motion planning problem for the nilpotent system.

Remark 3.4 It is also possible to produce piecewise controls u^i such that $S^{u^i}(T)$ exactly equals $\exp(q_i X_{I_i})$, as explained in [15]. Such a control avoids the computation of $\overline{q}_{i+1}, \ldots, \overline{q}_n$ at each step.

Of course the key point for this method is Assumption (A). It holds for instance in the following cases:

- the coordinates x in which the system is given, are the canonical coordinates of the second kind centered at $x^{\mathrm{initial}} = 0$ and associated with an adapted basis $X_{I_1}, \ldots, X_{I_n}$ at 0; in this case $\tilde{n} = n$ and $q(x) = x$;
- $(X_1, \ldots, X_m)$ is in canonical form in the coordinates x and $x^{\mathrm{initial}} = 0$; as noticed in Remark 2.7, this is a particular case of the previous one when one choose $\{I_1, \ldots, I_{\tilde{n}}\} = \mathscr{H}^r$.

There is however a general method to construct the mapping $x \mapsto q(x)$, and so to show that (A) holds for any real analytic nilpotent system. This method, that have been introduced in [21], works as follows.

We choose first a family of brackets $X_{I_1}, \ldots, X_{I_{\tilde{n}}}$ that generates the Lie algebra $Lie(X_1, \ldots, X_m)$ (which is finite dimensional since the system is nilpotent). In particular this family satisfy (i). For instance $I_1, \ldots, I_{\tilde{n}}$ may be chosen as elements of the P. Hall basis $\mathscr{H}^r$ (see Sect. 2.4.2).

The second step is to construct for every $x^{\mathrm{final}} \in U$ a $\tilde{n}$-tuple $(q_1, \ldots, q_{\tilde{n}})$ such that

$$x^{\mathrm{final}} = \exp(q_{\tilde{n}} X_{I_{\tilde{n}}}) \circ \cdots \circ \exp(q_1 X_{I_1})(x^{\mathrm{initial}}).$$

Let us introduce the extended control system,

$$\dot{x} = v_1 X_{I_1}(x) + \cdots + v_{\tilde{n}} X_{I_{\tilde{n}}}(x), \tag{3.28}$$

where $v = (v_1, \ldots, v_{\tilde{n}})$ is the control. It is easy to find a control v which steers the system (3.28) from x^{initial} to x^{final}. Indeed, we choose a C^1 path $\gamma : [0, T] \to \mathbb{R}^n$ connecting x^{initial} to x^{final} (a segment for example), and, for all $t \in [0, T]$, we express the tangent vector $\dot{\gamma}(t)$ as a linear combination of $X_{I_1}(t), \ldots, X_{I_{\tilde{n}}}(t)$. The coefficients of this combination are exactly $v_1(t), \ldots, v_{\tilde{n}}(t)$.

Denote by $S^v(t)$ the flow of the time-dependent vector field $\sum_i v_i X_{I_i}$. By construction $x^{\mathrm{final}} = S^v(T)(x^{\mathrm{initial}})$. Since $X_{I_1}, \ldots, X_{I_{\tilde{n}}}$ generate $Lie(X_1, \ldots, X_m)$, one can show that there exist $\tilde{n}$ functions $h_1^v, \ldots, h_{\tilde{n}}^v$ defined on $[0, T]$ such as

$$S^v(t) = \exp(h_{\tilde{n}}^v(t) X_{I_{\tilde{n}}}) \circ \cdots \circ \exp(h_1^v(t) X_{I_1}), \tag{3.29}$$

To compute these functions $h_i^v(\cdot)$, we replace $S^v(t)$ by its expression (3.29) in the differential equation,

$$\dot{S}^v(t) = \sum_{i=1}^{\tilde{n}} v_i(t) X_{I_i}(S^v(t)), \qquad S^v(0) = \mathrm{Id}.$$

It appears that, once expressed in terms of $h_1^v, \ldots, h_{\tilde{n}}^v$, this differential equation is a triangular system of the form,

$$\dot{h}_k^v = Q_k(h_1^v, \ldots, h_{k-1}^v, v_1, \ldots, v_k), \qquad k = 1, \ldots, \tilde{n}, \tag{3.30}$$

with $h_1(0) = \cdots = h_{\tilde{n}}(0) = 0$. We then compute the functions $h_1^v, \ldots, h_{\tilde{n}}^v$ by a direct integration. Since $x^{\mathrm{final}} = S^v(T)(x^{\mathrm{initial}})$, we obtain $q = q(x^{\mathrm{final}})$ by setting $q_i = h_i(T)$ for $i = 1, \ldots, \tilde{n}$. We have constructed in this way the mapping $x \mapsto q(x)$ of (ii), and hence ensured that Assumption (A) holds.

3.3 Method by Approximation

When the nonholonomic system is non nilpotentizable, we will look for an approximate solution to the motion planning problem rather than an exact one. The notion of first-order approximation for nonholonomic systems we have introduced in Chap. 2 suggests the use of a Newton type iterative method. Such a method works on the following scheme: solve first the motion planning problem for a first-order approximation of the system, and apply the so obtained control to the original system; iterate then the procedure from the resulting point. Taking this idea as a starting point, we present in this section a complete procedure for solving the motion planning problem. This procedure, which is inspired by the one in [16], has been introduced in [9].

3.3.1 Steering by Approximation

Let us first explain informally the principle of a method by approximation.

Given an initial point x^{initial} and a final point x^{final}, one first solves the motion planning problem for a nilpotent approximation of (3.1) at x^{final} by using one of the methods described in Sect. 3.2; then, one applies the resulting input $\hat{u}$ to (3.1) and iterates the procedure from the current point. If we use $\hat{\gamma}(t; p, u)$, $t \in [0, T]$, to denote the trajectory of the nilpotent approximation associated with the control u and starting at p, a local version of this algorithm is summarized in Algorithm 3.3, where d is the sub-Riemannian distance associated with (3.1) and e is a fixed positive real number.

Algorithm 3.3 Local Steering Algorithm

Require: $x^{\text{initial}}, x^{\text{final}}, e$
  $k := 0$;
  $x^k := x^{\text{initial}}$;
  while $d(x^k, x^{\text{final}}) > e$ **do**
    Compute $\hat{u}^k$ such that $x^{\text{final}} = \hat{\gamma}(T; x^k, \hat{u}^k)$;
    Set $x^{k+1} := \gamma(T; x^k, \hat{u}^k)$;
    $k := k + 1$;

The "while" loop in Algorithm 3.3 defines a mapping AppSteer : $(x^k, x^{\text{final}}) \mapsto \gamma(T; x^k, \hat{u}^k)$. The algorithm converges locally provided that AppSteer is *locally contractive* with respect to the distance d, i.e., for $x^{\text{final}} \in \Omega$, there exists $\varepsilon(x^{\text{final}}) > 0$ and $c(x^{\text{final}}) \in (0, 1)$ such that

$$d(x^{\text{final}}, \text{AppSteer}(x, x^{\text{final}})) \le c(x^{\text{final}})d(x^{\text{final}}, x), \tag{3.31}$$

for $x \in \Omega$ and $d(x^{\text{final}}, x) < \varepsilon(x^{\text{final}})$.

Assume now that we have a *uniformly locally contractive* mapping AppSteer on a connected compact set $K \subset \Omega$, i.e. that there exists $\varepsilon_K > 0$ and $c_K \in (0, 1)$ such that

$$d(x^{\text{final}}, \text{AppSteer}(x, x^{\text{final}})) \le c_K d(x^{\text{final}}, x), \tag{3.32}$$

for $x, x^{\text{final}} \in K$ and $d(x^{\text{final}}, x) < \varepsilon_K$. In this case the local algorithm above can be transformed into a global one (on K), for instance by using the following idea. Choose a path $\Gamma \subset K$ connecting x^{initial} to x^{final} and pick a finite sequence of intermediate goals $\{x_0^d = x^{\text{initial}}, x_1^d, \ldots, x_j^d = x^{\text{final}}\}$ on Γ such that $d(x_{i-1}^d, x_i^d) < \varepsilon_K/2$, $i = 0, \ldots, j$. It is then clear that the iterated application of a uniformly locally contractive $\text{AppSteer}(x^{i-1}, x_i^d)$ from the current state to the next subgoal (having set $x_i^d = x^{\text{final}}$ for $i \ge j$) yields a sequence x^i converging to x^{final}.

To turn the above idea into a practically efficient algorithm, two issues must be successfully addressed:

- construct a mapping AppSteer which is uniformly locally contractive;
- devise a globally convergent algorithm that do not use explicitly the "critical distance" ε_K, the knowledge of the latter being not available in practice.

3.3.2 Local Steering Method

In this section and the next one we will assume that the vectors fields $X_1, \ldots, X_m$ generate a free up to step r Lie algebra. We have seen in Sect. 3.1 that, up to a desingularization, we can always reduce the problem to that case. Recall that with that assumption every point $x \in \Omega$ is regular, the growth vector is constant on Ω, and the weights at every point are equal to the free weights, that is $w_j = \tilde{w}_j$ for $j = 1, \ldots, n$. Moreover the dimension n is equal to the dimension $\tilde{n}_r$ of $\mathscr{L}^r$.

Our aim now is to design a mapping AppSteer which is uniformly locally contractive, so that the corresponding steering Algorithm 3.3 is locally convergent. The first ingredient is the construction of a continuous nilpotent approximation, which itself requires the construction of a continuous varying system of privileged coordinates. For the latter we adapt the algebraic coordinates given in Sect. 2.1.2.

Construction of the nilpotent approximation $\mathscr{A}$

Denote by $x = (x_1, \ldots, x_n)$ the canonical coordinates in $\mathbb{R}^n$. For every point p in Ω, we construct the nilpotent approximation $\mathscr{A}(p)$ of $(X_1, \ldots, X_m)$ at p as follows.

(i) Take $\{X_{I_j} : I_j \in \mathscr{H}^r\}$.

(ii) Compute the affine change of coordinates $x \mapsto y = (y_1, \ldots, y_n)$ such that the new coordinates y satisfy $\partial_{y_j} = X_{I_j}(p)$, $j = 1, \ldots, n$.

(iii) Build the system of privileged coordinates $\widetilde{z} = (\widetilde{z}_1, \ldots, \widetilde{z}_n)$ by the following iterative formula, for $j = 1, \ldots, n$,

$$\widetilde{z}_j := y_j - \sum_{k=2}^{w_j - 1} h_k(y_1, \ldots, y_{j-1}), \tag{3.33}$$

where, for $k = 2, \ldots, w_j - 1$,

$$h_k(y_1, \ldots, y_{j-1}) = \sum_{\substack{|\alpha|=k \\ w(\alpha)<w_j}} X_{I_1}^{\alpha_1} \ldots X_{I_{j-1}}^{\alpha_{j-1}} \cdot (y_j - \sum_{q=2}^{k-1} h_q)(y)|_{y=0} \frac{y_1^{\alpha_1}}{\alpha_1!} \cdots \frac{y_{j-1}^{\alpha_{j-1}}}{\alpha_{j-1}!},$$

with $|\alpha| = \alpha_1 + \cdots + \alpha_n$.

(iv) For $i = 1, \ldots, m$, compute the Taylor expansion of $X_i(\widetilde{z})$ at 0, and express every vector field as a sum of vector fields which are homogeneous with respect to the weighted degree defined by the sequence $(w_j)_{j=1,\ldots,n}$:

$$X_i(\widetilde{z}) = X_i^{(-1)}(\widetilde{z}) + X_i^{(0)}(\widetilde{z}) + \cdots,$$

where we use $X_i^{(k)}(\widetilde{z})$ to denote the sum of all the terms of weighted degree equal to k.

(v) Define the vector fields $\widehat{X}_1^p, \ldots, \widehat{X}_m^p$ on Ω by $\widehat{X}_i^p := \widetilde{z}^* X_i^{(-1)}$ for $i = 1, \ldots, m$. Set $\mathscr{A}(p) := (\widehat{X}_1^p, \ldots, \widehat{X}_m^p)$.

(vi) For $j = 1, \ldots, n$, identify homogeneous polynomials Ψ_j of weighted degree equal to w_j such that, in the system of privileged coordinates $z = (z_1, \ldots, z_n)$ defined by
$$z_j = \widetilde{z}_j + \Psi_j(\widetilde{z}_1, \ldots, \widetilde{z}_{j-1}), \ j = 1, \ldots, n,$$

the family $(\widehat{X}_1^p, \ldots, \widehat{X}_m^p)$ is in the canonical form.

(vii) Define $\Phi(p, \cdot)$ as the mapping $x \mapsto z$.

The outputs of this algorithm are the mappings Φ and $\mathscr{A}$, which are respectively a continuously varying system of privileged coordinates and a continuous nilpotent approximation of $(X_1, \ldots, X_m)$ on Ω (see also Sect. 2.2.2).

Remark 3.5 The existence of the polynomials Ψ_j in Step (vi) is guaranteed by the same kind of arguments than the one in Step (v) of the Desingularization Algorithm, see Remark 2.8. The role of that step is to construct an approximated system $\mathscr{A}(p)$ that has always the same form in coordinates z, regardless of the vector fields $(X_1, \ldots, X_m)$ and of the approximation point $p \in \Omega$. The specificity of each system or each approximation point is then hidden in the change of coordinates Φ.

Now that a nilpotent approximation has been chosen, one must choose a way to control it. Let us denote by $\Lambda \subset \Omega \times \Omega$ the neighbourhood of the diagonal $\{(p, p), p \in \Omega\}$ where the mapping $(p, q) \mapsto \mathscr{A}(p)(q)$ is well-defined and continuous .

Definition 3.3 A *steering law* for $\mathscr{A}$ is a mapping which, with every pair $(x, p) \in \Gamma$, associates a control $\hat{u} \in L^1([0, T], \mathbb{R}^m)$, henceforth called a *steering control*, such that the trajectory $\hat{\gamma}(\cdot; x, \hat{u})$ of the *approximated system*,

$$\dot{x} = \sum_{i=1}^{m} \hat{u}_i \widehat{X}_i^p(x), \tag{3.34}$$

is defined on $[0, T]$ and satisfies $\hat{\gamma}(T; x, \hat{u}) = p$. In other words, $\hat{u}(\cdot)$ steers (3.34) from x to p.

A steering law of a nilpotent approximation is intended to be used as an approximate steering law for the original system. For that purpose, it is important to have a continuity property of the steering control: the closest are x and p, the smaller is the length of $\hat{u}$. We thus introduce the notion of *sub-optimality* (which is a sort of Lipschitz continuity of the steering law).

Definition 3.4 We say that a steering law for $\mathscr{A}$ is *sub-optimal* if there exists a constant $C_\ell > 0$ and a continuous positive function $\varepsilon_\ell(\cdot)$ such that, for any $(p, x) \in \Gamma$ with $d(p, x) < \varepsilon_\ell(p)$, the control $\hat{u}(\cdot)$ steering (3.34) from x to p satisfies:

$$\|\hat{u}\|_{L^1} \leq C_\ell \, \hat{d}_p(x, p) = C_\ell \, \hat{d}_p(\hat{\gamma}(0; x, \hat{u}), \hat{\gamma}(T; x, \hat{u})), \tag{3.35}$$

where $\hat{d}_p$ is the sub-Riemannian distance associated with $\mathscr{A}(p)$.

Recall that by definition, $\hat{d}_p(x, p)$ is the infimum of $\|u\|_{L^1}$ among all controls u steering (3.34) from x to p. As a consequence, sub-optimal steering laws always exist (and the constant C_ℓ can not be smaller than 1).

Looking for a steering law for $\mathscr{A}$ is much simpler in coordinates $z = \Phi(p, \cdot)$. Indeed, in these coordinates $\mathscr{A}(p)$ is in canonical form and so independent of p, and $z(p) = 0$. Hence all conditions are met in order to use the function $\mathrm{Steer}_{m,r}(\cdot)$

constructed in Sect. 3.2.2 for steering (3.34) from a point of coordinate z to p. Moreover one has $\hat{d}_p(x, p) = d_0(z(x), 0)$, where d_0 is the distance on $\mathbb{R}^{\tilde{n}}$ associated with the canonical form. As a direct consequence of Proposition 3.1 we then obtain the following result.

Corollary 3.1 *The mapping* $(x, p) \mapsto \mathrm{Steer}_{m,r}(\Phi(p, x))$ *is a sub-optimal steering law for* $\mathscr{A}$.

From now on, we assume that a sub-optimal steering law for $\mathscr{A}$ is given. We do not impose the choice of $\mathrm{Steer}_{m,r} \circ \Phi$ since there exist other solutions to the motion planning problem for nilpotent systems.

Definition 3.5 Given a steering law for $\mathscr{A}$, we define the mapping AppSteer from Γ to Ω as

$$\mathrm{AppSteer}(x, p) = \gamma(T; x, \hat{u}),$$

where $\hat{u}(\cdot)$ is the steering control of $\mathscr{A}(p)$ associated with (x, p).

Using a fixed point argument, the following result will ensure the local convergence of the local steering algorithm (Algorithm 3.3) based on the mapping AppSteer.

Proposition 3.2 *Let K be a compact subset of Ω. Assume that $\mathscr{A}$ is provided with a sub-optimal steering law. Then the associated mapping* AppSteer *is uniformly locally contractive on K, that is, there exists a constant $\varepsilon_K > 0$ such that, for every pair $(p, x) \in (K \times K) \cap \Gamma$ verifying $d(p, x) < \varepsilon_K$, there holds,*

$$d(p, \mathrm{AppSteer}(x, p)) \leq \frac{1}{2} d(p, x), \tag{3.36}$$

$$\|z(\mathrm{AppSteer}(x, p))\|_p \leq \frac{1}{2} \|z(x)\|_p, \tag{3.37}$$

where $z = \Phi(p, \cdot)$.

Note that the pseudo-norm $\|\cdot\|_p$ does not depend on $p \in K$ since the growth vector is constant on K.

Proof Since Φ is a continuous varying system of privileged coordinates and $\mathscr{A}$ a continuous nilpotent approximation, it follows from Theorem 2.3 that there exist continuous positive functions $C(\cdot)$ and $\varepsilon(\cdot)$ such that, for every pair $(x, p) \in \Omega \times \Omega$ with $d(x, p) < \varepsilon(p)$ and every control $u(\cdot)$ with $\|u\|_{L^1} < \varepsilon(p)$, there holds,

$$\|z(\gamma(T; x, u)) - z(\hat{\gamma}(T; x, u))\|_p \leq C(p) \max\left(\|z(x)\|_p, \|u\|_{L^1}\right) \|u\|_{L^1}^{1/r},$$

where r is the degree of nonholonomy at p, and $\hat{\gamma}(\cdot; x, u)$ is a trajectory of the nonholonomic system defined by $\mathscr{A}(p)$. Using the definition of a sub-optimal steering law and Inequality (2.13) in Theorem 2.3, one concludes easily. $\square$

Remark 3.6 Another way to construct a uniformly locally contractive mapping AppSteer is to extend the method described at the end of Sect. 3.2.3 to the situation where the system is not nilpotent. The difficulty in that case is that the Lie algebra $Lie(X_1, \ldots, X_m)$ is not finite-dimensional, implying that:

(i) the Campbell-Hausdorff formula contains an infinite number of terms, and so the procedure for constructing piecewise constant controls is no more finite;
(ii) the expression (3.29) is a composition of an infinite number of terms.

The approximation process consists then in keeping only the factors generated by brackets of order not greater than r in these formula. We can in this way define a mapping AppSteer which is locally contractive, see [21] for more details.

3.3.3 Global Steering Method

Let $K \subset \Omega$ be a connected compact set and x^{initial}, x^{final} be two points in K. We devise, under the assumptions of Proposition 3.2, an algorithm (Algorithm 3.3) which steers (3.1) from x^{initial} to a point arbitrarily close to x^{final}. That algorithm is a globalization of the local steering method Algorithm 3.3 but it does not require any a priori knowledge on the critical distance ε_K. This approach of the globalization is inspired by the trust-region methods in optimization (see for instance [3]).

Recall first that the family of vectors fields $(X_1, \ldots, X_m)$ is assumed to be free up to step r. As a consequence the weights $(w_1, \ldots, w_n)$, and so the pseudo-norm, are the same at any point $p \in \Omega$. We use the notation $\| \cdot \|_0$ for the pseudo-norm at any point of Ω. We introduce the parameterized path $t \mapsto \delta_t(x)$, which is defined by

$$\delta_t(x) = (t^{w_1} z_1(x), \ldots, t^{w_n} z_n(x)), \quad \text{for } x \in \Omega,$$

where $z = \Phi(x^{\text{final}}, \cdot)$. Note that δ_t is the (weighted) dilation in privileged coordinates at x^{final} with parameter t. In particular, $\|z(\delta_t(x))\|_0 = |t| \, \|z(x)\|_0$. We also define the function Subgoal as follows.

$$\begin{array}{|l|}
\hline
\text{Subgoal}(\overline{x}, \eta, j) \\
1.\ t_j := \max(0, 1 - \frac{j\eta}{\|z(\overline{x})\|_0}); \\
2.\ \text{Subgoal}(\overline{x}, \eta, j) := \delta_{t_j}(\overline{x}) \\
\hline
\end{array}$$

We note that the formula for generating t_j guarantees that

$$\|z(\text{Subgoal}(\overline{x}, \eta, j)) - z(\text{Subgoal}(\overline{x}, \eta, j - 1))\|_0 \leq \eta,$$

and that $x^d = x^{\text{final}}$ for j large enough.

The global algorithm is described in Algorithm 3.4 and used the mapping AppSteer as a sub-process. It steers (3.1) from x^{initial} to a point x such that $\|z(x)\|_0 \leq e$, where $e > 0$ is the desired precision (remind that $z = \Phi(x^{\text{final}}, \cdot)$, and so that $\|z(x^{\text{final}})\|_0 = 0$). Note that the Euclidean norm $\|\cdot\|$ is smaller than the pseudo-norm $\|\cdot\|_0$, hence the final point x of the algorithm satisfies $\|z(x) - z(x^{\text{final}}\| \leq e$.

Algorithm 3.4 Global $(x^{\text{initial}}, x^{\text{final}}, e, K, \text{AppSteer})$

1: $i := 0; \ j := 1;$

2: $x_i := x^{\text{initial}}; \ \overline{x} := x^{\text{initial}};$

3: $\eta := \|z(x^{\text{initial}})\|_0;$ {initial choice of the maximum step size}

4: **while** $\|z(x_i)\|_0 > e$ **do**

5:     $x^d := \text{Subgoal}\ (\overline{x}, \eta, j);$

6:     $x := \text{AppSteer}\ (x_i, x^d);$

7:     **if** $\|\Phi(x^d, x)\|_0 > \frac{1}{2}\|\Phi(x^d, x_i)\|_0$ **then** {if the system is not approaching the subgoal,}

8:       $\eta := \frac{\eta}{2};$ {reduce the maximum step size}

9:       $\overline{x} := x_i; \ j := 1;$ {change the path $\delta_{0,t}(\overline{x})$}

10:    **else**

11:       $i := i + 1; \ j := j + 1;$

12:       $x_i := x; \ x_i^d := x^d;$

13: **return** $x_i.$

Proposition 3.3 *Let $K \subset \Omega$ be a connected compact set equal to the closure of its interior. Assume that the mapping AppSteer is defined by a sub-optimal steering law of $\mathscr{A}$.*

Then, for any pair of points $(x^{\text{initial}}, x^{\text{final}}) \in K \times K$, Algorithm 3.4 terminates in a finite number of steps for any choice of the tolerance $e > 0$ provided that the sequences $(x_i)_{i \geq 0}$ and $(x_i^d)_{i \geq 0}$ both belong to K.

Remark 3.7 The last assumption of the proposition prevents the sequences $(x_i)_{i \geq 0}$ and $(x_i^d)_{i \geq 0}$ from accumulating on the boundary of the compact K. This assumption is of a purely numerical nature, and it can be removed either by adding suitable intermediate steps to Algorithm 3.4 as in [9], or by using numerical artifacts of probabilistic nature. Note also that if the points x^{initial} and x^{final} are far enough from the boundary of K, one can prove that the sequences $(x_i)_{i \geq 0}$ and $(x_i^d)_{i \geq 0}$ will remain in K.

Proof (of Proposition 3.3) Note first that, if the conditional statement of Line 7 is not true for every i greater than some i_0, then $x_i^d = x^{\text{final}}$ after a finite number of iterations. In this case, the error $\|z(x_i)\|_0$ is reduced at each iteration and the algorithm stops when it becomes smaller than the given tolerance e. This happens in particular if $d(x_i, x^d) < \varepsilon_K$ for all i greater than i_0 because condition (3.37) is verified. Another preliminary remark is that, due to the continuity of the distance d and of the function $\|z(\cdot)\|_0$, there exists $\overline{\eta} > 0$ such that, for every pair $(x_1, x_2) \in K \times K$, one has

$$\|z(x_1) - z(x_2)\|_0 < \overline{\eta} \implies d(x_1, x_2) < \frac{\varepsilon_K}{2}. \tag{3.38}$$

In the following, we will prove by induction that if, at some step i_0, one has $\eta < \overline{\eta}$, then, for all $i > i_0$,

$$d(x_{i-1}, x_i^d) < (1/2 + \cdots + (1/2)^{i-i_0})\varepsilon_K < \varepsilon_K.$$

We assume without loss of generality that $i_0 = 0$ and $\overline{x} = x_0$. For $i = 1$, by construction, $x^d = \text{Subgoal}(x_0, \eta, 1)$ and

$$\|z(x_0) - z(x^d)\|_0 \leq \eta < \overline{\eta}.$$

In view of (3.38), one obtains $d(x_0, x^d) < \varepsilon_K/2$, which implies by (3.37) that the conditional statement of Line 7 is not true. Therefore $x_1^d = x^d$ and $d(x_0, x_1^d) < \varepsilon_K/2$.

Assume now that for $i > 1$ one has

$$d(x_{i-2}, x_{i-1}^d) < (1/2 + \cdots + (1/2)^{i-1})\varepsilon_K. \tag{3.39}$$

The subgoal x_{i-1}^d is of the form $\text{Subgoal}(\overline{x}, \eta, j)$. Letting $x^d = \text{Subgoal}(\overline{x}, \eta, j+1)$, one has

$$d(x_{i-1}, x^d) \leq d(x_{i-1}, x_{i-1}^d) + d(x_{i-1}^d, x^d).$$

By construction, it is

$$\|z(x_{i-1}^d) - z(x^d)\|_0 \leq \eta < \overline{\eta},$$

which implies $d(x_{i-1}^d, x^d) < \varepsilon_K/2$. The induction hypothesis (3.39) implies that

$$d(x_{i-1}, x_{i-1}^d) \leq \frac{1}{2}d(x_{i-2}, x_{i-1}^d).$$

Finally, one gets

$$d(x_{i-1}, x^d) \leq \frac{1}{2}d(x_{i-2}, x_{i-1}^d) + d(x_{i-1}^d, x^d)$$

$$\leq (1/2 + \cdots + (1/2)^i)\varepsilon_K.$$

In view of (3.37), the conditional statement of Line 7 is not true, and so $x_i^d = x^d$. This ends the induction.

Notice that, at some step i, $\eta \geq \overline{\eta}$, the conditional statement of Line 7 could be false. In that case, η is decreased as in Line 8. The updating law of η guarantees that after a finite number of iterations of Line 8, there holds $\eta < \overline{\eta}$. This ends the proof. $\square$

3.3.4 A Complete Algorithm

We can now put together the elements developed in the section and present a global motion planning strategy. It is presented as an algorithmic procedure associated with a given nonholonomic system (3.1) defined on $\Omega \subset \mathbb{R}^n$ by m vector fields $X_1, \ldots, X_m$ which satisfy Chow's Condition. The required inputs are initial and final points x^{initial} and x^{final} belonging to Ω, a tolerance $e > 0$, and a compact connected set $K \subset \Omega$ (of appropriate size) equal to the closure of its interior which is a neighbourhood of both x^{initial} and x^{final}. For instance, K can be chosen to be a large enough compact tubular neighbourhood constructed around a curve joining x^{initial} and x^{final}.

Global Approximate Steering Method

Let $(x^{\text{initial}}, x^{\text{final}}, e, K)$ be given.

1. Build a decomposition (3.4) of K into a finite number of compact sets $K_i \subset \mathcal{V}_{\mathcal{J}_i}$, with $i = 1, \ldots, M$. Up to renumbering we assume $x^{\text{initial}} \in K_1$, $x^{\text{final}} \in K_{\bar{M}}$ for some integer $\bar{M} \leq M$, and $K_i \cap K_{i+1} \neq \emptyset$ for $i = 1, \ldots, \bar{M} - 1$. We can then choose a sequence $(x^i)_{i=1,\ldots,\bar{M}-1}$ such that $x^i \in K_i \cap K_{i+1}$ and we set $x^{\bar{M}} = x^{\text{final}}$.
2. Set $x := x^{\text{initial}}$.
3. For $i = 1, \ldots, \bar{M}$:

   a. Apply the Desingularization Algorithm at $p = x^i$ with $\mathcal{J} = \widetilde{\mathcal{J}}_i$ (see Sect. 2.4.2). The output is a system of coordinates z on $\mathcal{V}_{\mathcal{J}_i} \times \mathbb{R}^{\widetilde{n}_r - n}$, and m vector fields $\xi_1, \ldots, \xi_m$ on this domain.

   b. Define AppSteer as the mapping associated with the nilpotent approximation $\mathcal{A}$ of $(\xi_1, \ldots, \xi_m)$ on $\mathcal{V}_{\mathcal{J}_i} \times \mathbb{R}^{\widetilde{n}_r - n}$ and with its steering law $\text{Steer}_{m,r} \circ \Phi$ (see Sect. 3.3.2).

   c. Set $\widetilde{x}^{\text{initial}} := (x, 0)$, $\widetilde{x}^{\text{final}} := (x^i, 0)$. Choose $e_i > 0$ so that the set $\{y \in K_i : \|z(y) - z(x^i)\| \leq e_i\}$ is included in K_{i+1} if $i < \bar{M}$, and $e_i = e$ if $i = \bar{M}$. Choose also a closed Euclidean ball $\overline{B}(0, R)$ in $\mathbb{R}^{\widetilde{n}_r - n}$ of large radius R, and set $\widetilde{K}_i = K_i \times \overline{B}(0, R)$.

   d. Apply $\text{Global}(\widetilde{x}^{\text{initial}}, \widetilde{x}^{\text{final}}, e_i, \widetilde{K}_i, \text{AppSteer})$ to $(\xi_1, \ldots, \xi_m)$. The algorithm stops at a point $\widetilde{x}$ which is e_i-close to $\widetilde{x}^{\text{final}}$ (in the Euclidean norm associated with the coordinates z).

   e. Replace x by $\pi(\widetilde{x})$, where $\pi : \mathcal{V}_{\mathcal{J}_i} \times \mathbb{R}^{\widetilde{n}_r - n} \to \mathcal{V}_{\mathcal{J}_i}$ is the canonical projection, and return to step 3 with $i := i + 1$ if $i < \bar{M}$.

This method steers, in a finite number of steps, the nonholonomic system (3.1) from x^{initial} to a point $x \in K$ arbitrarily close to x^{final} as $e \to 0$.

Let us mention some advantages of this method. First, it may be easily adapted for stabilization tasks since it relies on an iterative procedure (see [26]). Second, taking advantage of the fact that the algorithm may be adapted to produce a C^1 input, one can address the motion planning problem for the dynamical extensions of nonholonomic systems, that is, systems for which the control is not u but its derivative. Finally, let us point out the modular nature of the method: one can propose other approaches to obtain uniformly contractive local methods (like the one of Remark 3.6), or replace $\text{Steer}_{m,r}(\cdot)$ by another control strategy for nilpotent systems.

3.4 Two Other Methods

Beside the method by approximation developed in the previous section we present with less details two other motion planning methods that reveal interesting characteristics of nonholonomic systems.

3.4.1 Path Approximation

Assume that the degree of nonholonomy of (3.1) is bounded by an integer r on the whole domain Ω and consider the elements $I_1, \ldots, I_{\widetilde{n}_r}$ of $\mathcal{H}^r$. For every $x \in \Omega$ the family $X_1(x), \ldots, X_{\widetilde{n}}(x)$ generates the whole $\mathbb{R}^n$. It follows that any smooth curve in Ω is a trajectory of the extended system defined as,

$$\dot{x} = v_1 X_{I_1}(x) + \cdots + v_{\widetilde{n}_r} X_{I_{\widetilde{n}_r}}(x), \quad x \in \Omega, \tag{3.40}$$

where $v = (v_1, \ldots, v_{\widetilde{n}_r})$ is the control.

The idea of the path approximation method is to construct trajectories of the nonholonomic system (3.1) approaching the ones of the extended system. More precisely, given a trajectory γ of the extended system, we will explicitly construct a sequence $u^j = (u_1^j, \ldots, u_m^j)$ of controls such that the corresponding trajectories of (3.1) converge uniformly in time to γ when j tends to infinity. This result gives an approximate solution to the motion planning problem: choosing any smooth curve γ (which is also a trajectory of the extended system) that joins x^{initial} to x^{final}, the control u^j constructed above will steer (3.1) arbitrarily close to x^{final} for j large enough.

The main advantage of the construction is its universal character: it is independent of the particular values taken by the vector fields, and depends only on the structure of the Lie brackets. Therefore, the natural environment to present this construction is that of the free Lie algebra. However, this abstraction makes thinks difficult to understand, so we have chosen to describe the method on a simple example. We refer the reader to the papers of Sussmann and Liu for a complete presentation (first appearance in [29], all technical details in [19], and a simplified but more affordable presentation in [30]).

Consider then the case of a system with two inputs in $\mathbb{R}^5$, that is,

$$\dot{x} = u_1 X_1(x) + u_2 X_2(x), \qquad x \in \mathbb{R}^5, \tag{3.41}$$

whose degree of nonholonomy is never greater than 3. The elements $I_1, \ldots, I_{\tilde{n}_r}$ of $\mathscr{H}^3$ are $I_1 = 1$, $I_2 = 2$, $I_3 = [1, 2]$, $I_4 = [1, [1, 2]]$, and $I_5 = [2, [1, 2]]$. Fix two points x^{initial}, x^{final} in $\mathbb{R}^5$ and choose a C^1 curve $\gamma : [0, 1] \to \mathbb{R}^5$ such that $\gamma(0) = x^{\text{initial}}$ and $\gamma(1) = x^{\text{final}}$. There exist smooth real-valued functions $v_1, \ldots, v_5$ defined on $[0, 1]$ such that $t \to \gamma(t)$ is the solution of the Cauchy problem

$$\dot{x} = v_1(t) X_{I_1}(x) + \cdots + v_5(t) X_{I_5}(x), \qquad x(0) = x^{\text{initial}}.$$

We seek to build a control sequence $(u^j)_{j \in \mathbb{N}}$ such that the sequence of trajectories of (3.41) associated with $(u^j)_{j \in \mathbb{N}}$ converges uniformly in time to γ when j tends to infinity.

As in Sect. 3.2.2, we use oscillating controls that are linear combinations of sinusoidal functions with carefully chosen frequencies. Consider three groups of real-valued frequencies $\Omega_k = \{\omega_{k,1}, \omega_{k,2}\}$, for $k = 1, 2, 3$, and assume that the $\omega_{k,l}$ satisfy the following conditions:

(i) $\omega_{1,1} + \omega_{1,2} = 0$, $2\omega_{2,1} + \omega_{2,2} = 0$, $\omega_{3,1} + 2\omega_{3,2} = 0$;
(ii) for any integer α_k in $[-2, 2]$ and any $\omega_k \in \Omega_k, k = 1, 2, 3$, one has,

$$\alpha_1 \omega_1 + \alpha_2 \omega_2 + \alpha_3 \omega_3 = 0 \implies \alpha_1 = \alpha_2 = \alpha_3 = 0.$$

Now, fix some C^1 functions $\eta_1, \eta_2, \eta_{k,l}, k = 1, 2, 3$ and $l = 1, 2$. We define the sequence of controls $u^j = (u_1^j, u_2^j)$, $j \in \mathbb{N}$, by

$$u_1^j(t) = \eta_1(t) + u_{1,1}^j(t) + u_{2,1}^j(t) + u_{3,1}^j(t),$$
$$u_2^j(t) = \eta_2(t) + u_{1,2}^j(t) + u_{2,2}^j(t) + u_{3,2}^j(t),$$

where

$$u_{1,1}^j(t) = j^{\frac{1}{2}} \eta_{1,1}(t) \sin \omega_{1,1} jt, \qquad u_{1,2}^j = j^{\frac{1}{2}} \eta_{1,2}(t) \cos \omega_{1,2} jt, \tag{3.42}$$

$$u_{2,1}^j(t) = j^{\frac{2}{3}} \eta_{2,1}(t) \cos \omega_{2,1} jt, \qquad u_{2,2}^j = j^{\frac{2}{3}} \eta_{2,2}(t) \cos \omega_{2,2} jt, \tag{3.43}$$

$$u_{3,1}^j(t) = j^{\frac{2}{3}} \eta_{3,1}(t) \cos \omega_{3,1} jt, \qquad u_{3,2}^j = j^{\frac{2}{3}} \eta_{3,2}(t) \cos \omega_{3,2} jt. \tag{3.44}$$

Then the sequence of trajectories of the system (3.41) associated with the sequence of controls u^j converges uniformly to the solution of

$$\dot{x} = \eta_1 X_{I_1}(x) + \eta_2 X_{I_2}(x) - \frac{\eta_{1,1}\eta_{1,2}}{2\omega_{1,1}} X_{I_3}(x) - \frac{\eta_{2,1}^2\eta_{2,2}}{8\omega_{2,1}^2} X_{I_4}(x) - \frac{\eta_{3,1}\eta_{3,2}^2}{4\omega_{3,1}\omega_{3,2}} X_{I_5}(x).$$

As a consequence, by choosing the functions $\eta_1, \eta_2, \eta_{k,l}$ such that

$$\eta_1(t) = v_1(t), \quad \eta_2(t) = v_2(t), \quad -\frac{\eta_{1,1}(t)\eta_{1,2}(t)}{2\omega_{1,1}} = v_3(t),$$

$$-\frac{\eta_{2,1}^2(t)\eta_{2,2}(t)}{8\omega_{2,1}^2} = v_4(t), \quad -\frac{\eta_{3,1}(t)\eta_{3,2}^2}{4\omega_{3,1}\omega_{3,2}} = v_5(t),$$

the sequence of controls $(u^j)_{j\in\mathbb{N}}$ produces a sequence of trajectories of (3.41) that converge uniformly to γ. This ends the construction.

It is important to notice the role of the conditions (i) and (ii). For instance, the resonance condition (i) ensures that a control such as (3.42) does a motion in the direction X_3, whereas the independence condition (ii) guarantees that the controls (3.43) and (3.44) do not produce any displacement in that direction. Thus these conditions play exactly the same role as (3.17) and (3.18) do respectively for steering nilpotent systems.

However there is a huge difference between both strategies. Indeed, the path approximation requires to use sinusoidal controls whose frequencies tend to infinity in order to ensure that the final state converges to the goal, even if the system under consideration is nilpotent. This makes the method hardly usable in practice for most applications.

3.4.2 Continuation Method

Fix a point $p \in \Omega$ and a time $T > 0$. We will consider only L^2 control functions, so we define the set of admissible controls as $\mathcal{U} = L^2([0, T], \mathbb{R}^m)$. The *end-point mapping* at p in time T is defined by

$$E_{p,T} : v \in \mathcal{U} \mapsto \gamma(T; p, v).$$

Using this mapping, the motion planning problem may be restated as follows: given a pair $(x^{\text{initial}}, x^{\text{final}}) \in \Omega \times \Omega$, find a control $u \in \mathcal{U}$ such that

$$E_{x^{\text{initial}},T}(u) = x^{\text{final}}.$$

Thus, we want to invert the application $E_{x^{\text{initial}},T}$, or more precisely, we are looking for a right inverse of $E_{x^{\text{initial}},T}$ because it is surjective (the system is controllable), but it is not injective (there are several controls u such that $E_{x^{\text{initial}},T}(u) = x^{\text{final}}$). We proceed as follows.

Take any control $u^0 \in \mathcal{U}$. The idea is to build a path in $\mathcal{U}$ that allows to go from u^0 to a control u^1 satisfying $E_{x^{\mathrm{initial}},T}(u^1) = x^{\mathrm{final}}$. Set $x^0 = E_{x^{\mathrm{initial}},T}(u^0)$, and choose a path $\pi : [0, 1] \to \Omega$ such that $\pi(0) = x^0$ and $\pi(1) = x^{\mathrm{final}}$. We seek a path $\Pi : [0, 1] \to \mathcal{U}$ such that, for almost every $s \in [0, 1]$,

$$E_{x^{\mathrm{initial}},T}(\Pi(s)) = \pi(s).$$

Taking the derivative with respect to s, we obtain

$$dE_{x^{\mathrm{initial}},T}(\Pi(s)) \cdot \frac{d\Pi}{ds}(s) = \frac{d\pi}{ds}(s), \tag{3.45}$$

where $dE_{x^{\mathrm{initial}},T}(v)$ denotes the differential of $E_{x^{\mathrm{initial}},T}$ at $v \in \mathcal{U}$.

If $dE_{x^{\mathrm{initial}},T}(\Pi(s))$ is of full rank, it admits right inverses. Choose one of them (for instance the Moore-Penrose pseudo-inverse), and denote it by $P(\Pi(s))$. We can form a differential equation,

$$\frac{d\Pi}{ds}(s) = P\big(\Pi(s)\big) \cdot \frac{d\pi}{ds}(s), \tag{3.46}$$

whose solutions satisfy (3.45). As a consequence, the solution of the Cauchy problem,

$$\begin{cases} \dfrac{d\Pi}{ds}(s) = P\big(\Pi(s)\big) \cdot \dfrac{d\pi}{ds}(s), \\ \Pi(0) \;\;\; = u^0, \end{cases} \tag{3.47}$$

is by construction a solution of (3.45), provided it is defined on the whole interval $[0, 1]$, and its final value $u^1 = \Pi(1)$ is a solution of the motion planning problem between x^{initial} and x^{final}. It can be obtained through a numerical integration of (3.47) on $[0, 1]$ using for example an Euler scheme.

Hence we are led to study the lifting equation (3.46) as an ordinary differential equation on $\mathcal{U}$. In order to use this equation to solve the motion planning problem, we have to guarantee two conditions:

(a) non degeneracy: the path π must be chosen so that $dE_{x^{\mathrm{initial}},T}(\Pi(s))$ is of full rank for every $s \in [0, 1]$;

(b) non explosion: for any initial choice u^0, the solution of the Cauchy problem (3.47) must be defined on the whole interval $[0, 1]$.

Point *(a)* ensures the existence of $P(\Pi(s))$ for all $s \in [0, 1]$, and so that (3.46) is well-defined. Point *(b)* ensures the existence of $\Pi(1)$. Note that the end-point mapping is smooth, so we always have local existence and uniqueness of the solution of (3.47).

Both conditions are very difficult to guarantee. For the first point one has to deal with the *singular controls*, which are the critical points of the end-point mapping. A way to ensure *(a)* would be to require that the path π avoids the set of the critical values of the end-point mapping. However there is no general characterization of this set, we do not even know if it is of zero measure (that is, if the end-point mapping satisfies a Sard type property). We refer to [28] for a discussion on singular controls.

Point *(b)* consists in proving the global existence of the solutions of a highly nonlinear differential equation, which is particularly difficult in the general case. It is related to *(a)* since the interval of definition of the solutions depends on the domain of definition of the differential equation itself. Results exist only under very restrictive assumptions [6, 8].

Thus from a theoretical point of view, the validity of the continuation method is not settled. In particular a proof of convergence in a general setting seems to be very difficult to obtain. However the numerical implementation of this method (or of some of its variants) for concrete examples often provide rapidly a very satisfactory approximate solution to the motion planning problem. It is then worth to consider this method in practical problems, and to further investigate its mathematical setting.

3.5 An Overview of the Motion Planning Algorithms

To conclude this section, let us give a brief overview of the existing solutions to the motion planning problem. Recall that such a solution is an algorithm whose inputs are the points $(x^{\text{initial}}, x^{\text{final}})$, and whose output is a control u steering the system from x^{initial} to x^{final}, or to a point arbitrarily close to x^{final}. Several criteria may be introduced to judge such a procedure, we propose the following ones.

1. Generality: to which class of nonholonomic systems applies the algorithm? The larger the class is, the more general the algorithm will be. The most general algorithms apply to systems satisfying the sole Chow Condition.
2. Global character of the algorithm: does the algorithm produce a steering control for every pair of points in Ω? Or just for points close enough one to each other? Note that the core of many algorithms consists in a local procedure and turning the latter into a global one may be a serious obstacle to overcome.
3. Convergence: does a mathematical proof of convergence exist?
4. Usefulness for practical applications:

    - is the dimension n of the state drastically limited?
    - is the procedure robust with respect to the dynamics?
    - does it produce "nice" trajectories, e.g. smooth enough, without cusps nor large oscillations?
    - does it have a good local behaviour, i.e., is it possible to reduce the state space Ω to any smaller open and connected subset of $\mathbb{R}^n$?

5. Efficiency of the numerical implementations.

Let us present now the existing solutions to the nonholonomic motion planning problem at the light of the aforementioned criteria. We restrict our overview to the algorithms that present a sufficient degree of generality. In particular we do not mention algorithms applying to very specific systems, or to systems in small dimension (typically, there exist several particular methods in dimension 3).

At first, in the case of specific classes of nonholonomic systems (i.e. where more is known than the sole Chow Condition), effective techniques have been proposed, among which a Lie bracket method for steering nilpotentizable systems (see [18, 21] and Sect. 3.2.3), sinusoidal controls for chained systems (see [24] and Sect. 3.2.1), averaging techniques for left-invariant systems defined on a Lie group (see [4, 20]), and a trajectory generation method for flat systems (see [13]). Depending on the applications, these methods turn out to be extremely efficient, especially when the system to be steered is shown to be flat with an explicit flat output.

However, the class of systems considered previously is rather restrictive. For instance for 2-input nonholonomic systems (i.e. $m = 2$), under suitable regularity assumptions, a flat system is feedback equivalent to a chained system (cf. [23, 25]), which is a non generic property among nonholonomic systems as soon as the dimension of the state space is larger than 4. As already mentioned in Sect. 3.2, the same property of non genericity holds for nilpotentizable systems. Moreover, there exist standard nonholonomic systems whose kinematic model does not fall into any of the aforementioned categories. For instance, mobile robots with more than one trailer cannot be transformed in chained-form unless each trailer is hinged to the midpoint of the previous wheel axle, an unusual situation in real vehicles. Another similar example is the rolling-body problem: even the simplest model in this category, the so-called plate-ball system, does not allow any chained-form transformation and is not flat.

Regarding general nonholonomic systems, various steering techniques have been proposed in the literature. Let us first mention the generic loop method, presented in [31]. It is based on a local deformation procedure but requires an a priori estimate of some "critical distance" which is an unknown parameter in practice. That fact translates into a severe drawback for constructing a globally valid algorithm. The path approximation of [19, 29], which uses unbounded sequence of sinusoids, is also worth noticed (see Sect. 3.4.1). It is however not adapted to most of the practical motion planning tasks, since it relies on a limit process of highly oscillating inputs. A series of papers ([2, 14] and references therein) presents another method of path approximation based on optimal control but it applies only for certain classes of nonholonomic systems.

The continuation method of [8, 33] (presented in Sect. 3.4.2) belongs to the class of Newton type methods. Proving its convergence amounts to show the global existence for the solution of a non linear differential equation, which relies on handling the singular controls of the nonholonomic system. That latter issue turns out to be a hard one, see [10, 11] for instance. This is why, in the current state of knowledge, the continuation method can be proved to converge only under restrictive assumptions (see [6, 7, 12]). However the numerical implementations of this method seem to be very efficient in practice [1, 27, 34].

The most complete methods, with respect to the criteria defined previously, are iterative methods based on exact solution to the motion planning problem for nilpotent systems. The first one has been introduced in [18] and improved in [21]. It is a completely general method that suffers however from important limitation for practical use. First, the method is actually a local one, since its globalization requires the knowledge of a critical distance, which is not available in practice. Second, either the resulting trajectories in [21] contain a large number of cusps (exponential with respect to the degree of nonholonomy), or the computation of the steering control in [18] requires the inversion of a system of algebraic equations. The latter turns out to be numerically intractable as soon as the dimension of the state is larger than six. The second iterative method, introduced in [16] and developed in [9], is based on the notion of nilpotent approximation and has been detailed in Sect. 3.3. This method meets all our criteria except the last one. It has indeed not been fully implemented up to now, so its numerical efficiency remains to be seen.

References

1. Alouges, F., Chitour, Y., Long, R.: A motion-planning algorithm for the rolling-body problem. IEEE Trans. Robot. **26**(5), 827–836 (2010)
2. Boizot, Nicolas, Gauthier, Jean-Paul: Motion planning for kinematic systems. IEEE Trans. Automat. Control **58**(6), 1430–1442 (2013)
3. Bonnans, J.F., Gilbert, J-Ch., Lemaréchal, C., Sagastizábal, C.: Numerical Optimization-Theoretical and Practical Aspects, 2nd edn. Springer, Berlin (2006)
4. Bullo, F., Lewis, A., Leonard, N.: Controllability and motion algorithms for underactuated lagrangian systems on lie groups. IEEE Trans. Autom. Control **45**(8), 1437–1454 (2000)
5. Canny, J.F.: The Complexity of Robot Motion Planning. MIT Press, Cambridge, MA (1988)
6. Chelouah, A., Chitour, Y.: On the motion planning of rolling surfaces. Forum Mathematicum **15**, 727–758 (2003)
7. Chitour, Y.: Path planning on compact Lie groups using a continuation method. Syst. and Cont. Lett. **45**, 383–392 (2002)
8. Chitour, Y.: A continuation method for motion-planning problems. ESAIM Control Optimisation Calc. Var. **12**(01), 139–168 (2006)
9. Chitour, Y., Jean, F., Long, R.: A global steering method for nonholonomic systems. J. Differ. Equ. **254**, 1903–1956 (2013)
10. Chitour, Y., Jean, F., Trélat, E.: Genericity results for singular curves. J. Differ. Geom. **73**(1), 45–73 (2006)
11. Chitour, Y., Jean, F., Trélat, E.: Singular trajectories of control-affine systems. SIAM J. Control Optim. **47**(2), 1078–1095 (2008)
12. Chitour, Y, Sussmann, H.J.: Essays On Mathematical Robotics. In: Line-integral estimates and motion planning using a continuation method. IMA Vol. Math. Appl. 104. Springer, NY (1998)
13. Fliess, M., Lévine, J., Martin, P., Rouchon, P.: Flatness and defect of non-linear systems: introductory theory and examples. Int. J. Control **61**, 1327–1361 (1995)
14. Gauthier, J., Jakubczyk, B., Zakalyukin, V.: Motion planning and fastly oscillating controls. SIAM J. Control Optim. **48**(5), 3433–3448 (2010)
15. Jean, F., Koseleff, P.-V.: Elementary approximation of exponentials of Lie polynomials. In: Mora, T., Mattson, H. (eds) Proceedings 12th AAECC Symposium, Lecture Notes in Computer Science, vol. 1255, pp. 174–188. Springer, Berlin (1997)
16. Jean, F., Oriolo, G., Vendittelli, M.: A globally convergent steering algorithm for regular nonholonomic systems. In: Proceedings of 44th IEEE CDC-ECC'05, Sevilla, Spain (2005)
17. Kawski, Matthias: Nilpotent lie algebras of vectorfields. Journal fr die reine und angewandte Mathematik **388**, 1–17 (1988)

18. Lafferriere, G.: A general strategy for computing steering controls of systems without drift. In: 30th IEEE Conference on Decision and Control, Brighton, England (1991)
19. Liu, W.: An approximation algorithm for nonholonomic systems. SIAM J. Control Optim. **35**(4), 1328–1365 (1997)
20. Leonard, N., Krishnaprasad, P.S.: Motion control of drift-free, left-invariant systems on lie groups. IEEE Trans. Autom. Control **40**(9), 1539–1554 (1995)
21. Lafferriere, G., Sussmann, H.J: A differential geometric approach to motion planning. In: Li, Z., Canny, J.F. (eds) Nonholonomic Motion Planning. Kluwer (1992)
22. Laumond, J.-P., Sekhavat, S., Lamiraux, F.: Guidelines in nonholonomic motion planning for mobile robots. In: Laumond, J.-P. (ed.) Robot Motion Planning and Control, Lecture Notes in Information and Control Sciences, vol. 229. Springer, Berlin (1998)
23. Martin, P., Murray, R.M., Rouchon, P.: Flat systems: open problems, infinite dimensional extension, symmetries and catalog. In: Advances in the Control of Nonlinear Systems, Lecture Notes in Control and Information Sciences, pp. 33–57. Springer, Berlin (2001)
24. Murray, R.M., Sastry, S.S.: Nonholonomic motion planning: steering using sinusoids. IEEE Trans. Autom. Control **38**(5), 700–716 (1993)
25. Murray, R.M.: Nilpotent bases for a class of nonintegrable distributions with applications to trajectory generation for nonholonomic systems. Math. Control Signals Syst. **7**, 58–75 (1994)
26. Oriolo, G., Vendittelli, M.: A framework for the stabilization of general nonholonomic systems with an application to the plate-ball mechanism. IEEE Trans. Robot. **21**(2), 162–175 (2005)
27. Popa, Dan O., Wen, John T.: Singularity computation for iterative control of nonlinear affine systems. Asian J. Control **2**(2), 57–75 (2000)
28. Rifford, L.: Sub-Riemannian Geometry and Optimal Transport. SpringerBriefs in Mathematics. Springer International Publishing (2014)
29. Sussmann, H.J., Liu, W.: Limits of highly oscillatory controls and the approximation of general paths by admissible trajectories. In: 30th IEEE Conference on Decision and Control (1991)
30. Sussmann, H.J., Liu, W.: Lie bracket extensions and averaging: the single-bracket case. In: Li, Z., Canny, J.F. (eds) Nonholonomic Motion Planning, pp. 109–147. Kluwer Academic Publishers (1993)
31. Sontag, E.D.: Control of systems without drift via generic loops. IEEE Trans. Automat. Control **40**(7), 1210–1219 (1995)
32. Schwartz, J.T., Sharir, M.: On the 'piano movers' problem II: general techniques for computing topological properties of real algebraic manifolds. Adv. Appl. Math. **4**, 298–351 (1983)
33. Sussmann, H.J.: A continuation method for nonholonomic path-finding problems. In: Proceedings of 32nd IEEE CDC, pp. 2718–2723 (1993)
34. Tchoń, K., Jakubiak, J., Małek, Ł.: Motion planning of nonholonomic systems with dynamics. In: Kecskemthy, A., Mller, A. (eds) Computational Kinematics, pp. 125–132. Springer, Berlin (2009)

Appendix A
Composition of Flows of Vector Fields

This chapter is dedicated to the proof of Campbell-Hausdorff type formulas for flows of vector fields. The result in Sect. A.1 has been used in Sect. 1.4, the one in Sect. A.2 will be necessary for the next appendix.

Let U be an open subset of $\mathbb{R}^n$ and $VF(U)$ the set of smooth vector fields on U. Given a vector field $X \in VF(U)$, we denote its flow by $\exp(tX)$.

A.1 Campbell-Hausdorff Formula for Flows

We will need in this chapter the Campbell-Hausdorff formula which we recall briefly here (for a more detailed presentation see for instance [2, Chap. II]). Let x and y be two non commutative unknowns, and $[x, y] = xy - yx$ their commutator, also denoted by $[x, y] = (\mathrm{ad}x)y$. The length of an iterated commutator $(\mathrm{ad}x_1)\dots(\mathrm{ad}x_{k-1})x_k$, where each $x_1, \dots, x_k$ equals x or y, is defined to be the number of occurrences k of x and y. Define also e^x and e^y to be the series $\sum_{k\geq 0} \frac{x^k}{k!}$ and $\sum_{k\geq 0} \frac{y^k}{k!}$. Then we have $e^x e^y = e^{H(x,y)}$ in the sense of formal power series, where

$$H(x, y) = x + y + \frac{1}{2}[x, y] + R(x, y), \tag{A.1}$$

and $R(x, y)$ is a series whose terms are linear combination of iterated commutators of x and y of length greater than 2. For an integer N we denote by $H_N(x, y)$ the partial sum of $H(x, y)$ containing only iterated commutators of length not greater than N. In particular, $H_1 = x + y$ and $H_2 = x + y + \frac{1}{2}[x, y]$.

Consider now two vector fields $X, Y \in VF(U)$. Given $t \in \mathbb{R}$ and an integer N, $H_N(tY, tX)$ is a smooth vector field on U which writes as $\sum_{i=1}^{N} t^i Y_i$, where the vector fields $Y_1, \dots, Y_N$ belong to the Lie algebra generated by X and Y.

© The Author(s) 2014

F. Jean, *Control of Nonholonomic Systems: From Sub-Riemannian Geometry to Motion Planning*, SpringerBriefs in Mathematics, DOI 10.1007/978-3-319-08690-3

Lemma A.1 *For any $p \in M$, there exist positive constants δ and C such that $|t| < \delta$ implies*

$$\|\exp(tX) \circ \exp(tY)(p) - \exp(H_N(tY, tX))(p)\| \leq C|t|^{N+1}.$$

Proof Set $\psi(t) = \exp(tX) \circ \exp(tY)(p)$, which is a function defined and C^∞ in a neighbourhood of $0 \in \mathbb{R}$, and let $(x_1, \ldots, x_n)$ be a system of local coordinates on a neighbourhood of p in U. We will compute the Taylor expansion of every component $x_i(\psi(t))$, for $i = 1, \ldots, n$. To do this, we introduce the function $\phi(t, s) = \exp(tX) \circ \exp(sY)(p)$, so that $\psi(t) = \phi(t, t)$, and we compute the partial derivatives of $x_i \circ \phi$ at $0 \in \mathbb{R}^2$. We have:

$$\frac{\partial x_i \circ \phi}{\partial t}(t, s) = \frac{d}{dt}\left[x_i \circ \exp(tX)\right](\exp(sY)(p)) = Xx_i(\phi(t, s)),$$

where Xx_i denotes the Lie derivative of x_i along X. Repeating this computation, we obtain for any integer k,

$$\frac{\partial^k x_i \circ \phi}{\partial t^k}(t, s) = X^k x_i(\phi(t, s)).$$

In the same way, we have:

$$\frac{\partial^{k+l} x_i \circ \phi}{\partial s^l \partial t^k}(0, 0) = \frac{\partial^l}{\partial s^l} \frac{\partial^k x_i \circ \phi}{\partial t^k}(0, s)\Big|_{s=0}$$

$$= \frac{\partial^l}{\partial s^l}\left[X^k x_i(\exp(sY)(p))\right]\Big|_{s=0} = Y^l X^k x_i(p).$$

We then deduce that the formal Taylor series of $x_i(\psi(t)) = x_i(\phi(t, t))$ at 0 is

$$\sum_{k,l \geq 0} \frac{t^{k+l}}{k!l!} Y^l X^k x_i(p) = \left[\sum_{l \geq 0} \frac{t^l}{l!} Y^l\right]\left[\sum_{k \geq 0} \frac{t^k}{k!} X^k\right] x_i(p),$$

where X and Y are considered as derivation operators. From the Campbell-Hausdorff formula, the product of the formal series $e^{tY} = \sum_{l \geq 0} \frac{t^l}{l!} Y^l$ with $e^{tX} = \sum_{k \geq 0} \frac{t^k}{k!} X^k$ is equal to the series $e^{H(tY, tX)}$. As a consequence, the Taylor expansion of $x_i(\psi(t))$ up to degree N is given by the terms of degree $\leq N$ in the series $e^{H(tY, tX)} x_i(p)$, which coincide with the terms of degree $\leq N$ in the series $e^{H_N(tY, tX)} x_i(p)$.

On the other hand, it results from Lemma A.2 below that $e^{H_N(tY, tX)} x_i(p)$ is the Taylor series at 0 of the function $t \mapsto x_i \circ \exp(H_N(tY, tX))(p)$. Thus

$$x_i \circ \psi(t) - x_i \circ \exp(H_N(tY, tX))(p) = O(|t|^{N+1})$$

for every coordinate x_i. $\qquad\square$

Lemma A.2 *Let $Y_1, \ldots, Y_\ell$ be vector fields on U, $f: U \to \mathbb{R}$ a smooth function, and $p \in U$. The formal Taylor series at $0 \in \mathbb{R}^\ell$ of the function $(z_1, \ldots, z_\ell) \mapsto f(\exp(\sum_i z_i Y_i)(p))$ is given by*

$$\sum_{k \geq 0} \frac{1}{k!} \Big(\sum_i z_i Y_i\Big)^k f(p) = e^{\sum_i z_i Y_i} f(p).$$

The formal Taylor series at $0 \in \mathbb{R}$ of the function $t \mapsto f(\exp(Y(t))(p))$, where $Y(t) = \sum_{i=1}^\ell t^i Y_i$, is given by

$$\sum_{k \geq 0} \frac{1}{k!} Y(t)^k f(p) = e^{Y(t)} f(p). \tag{A.2}$$

Proof The second statement is obviously a consequence of the first one, so it is sufficient to prove the latter. We introduce the functions $g(z) = f(\exp(\sum_i z_i Y_i)(p))$ and $G(t, z) = g(tz)$ which are well-defined and smooth on a neighbourhood of 0 in $\mathbb{R}^\ell$, respectively $\mathbb{R} \times \mathbb{R}^\ell$. We are looking for the Taylor series of g at 0.

Since $G(t, z) = f(\exp(t \sum_i z_i Y_i)(p))$, we have, for any integer $k \geq 0$,

$$\frac{\partial^k G}{\partial t^k}(0, z) = \Big(\sum_i z_i Y_i\Big)^k f(p).$$

On the other hand $G(t, z) = g(tz)$, and hence the previous derivative can also be computed as

$$\frac{\partial^k G}{\partial t^k}(0, z) = \sum_{\alpha_1 + \cdots + \alpha_\ell = k} \frac{k!}{\alpha_1! \cdots \alpha_\ell!} z_1^{\alpha_1} \cdots z_\ell^{\alpha_\ell} \frac{\partial^k g}{\partial z_1^{\alpha_1} \cdots \partial z_\ell^{\alpha_\ell}}(0).$$

Combining both expressions, we obtain

$$\sum_{\alpha_1 + \cdots + \alpha_\ell = k} \frac{z_1^{\alpha_1} \cdots z_\ell^{\alpha_\ell}}{\alpha_1! \cdots \alpha_\ell!} \frac{\partial^k g}{\partial z_1^{\alpha_1} \cdots \partial z_\ell^{\alpha_\ell}}(0) = \frac{1}{k!} \Big(\sum_i z_i Y_i\Big)^k f(p),$$

and the lemma follows. $\qquad\square$

Lemma A.1 can be extended in two ways. First, since the vector fields and their flows are smooth on U, the estimate holds uniformly with respect to p. Second, by Lemma A.2, the vector fields tX and tY may be replaced by the one-parameter families of vector fields $X(t) = tX_1 + \cdots + t^k X_k$ and $Y(t) = tY_1 + \cdots + t^\ell Y_\ell$, where $X_1, \ldots, X_k$ and $Y_1, \ldots, Y_\ell$ are vector fields on U. As an example, $H_N(tY, tX)$ is of this form. To summarize, a slight change in the proof of Lemma A.1 actually shows the following result.

Corollary A.1 *Let $K \subset U$ be a compact. There exist two positive constants δ, C such that, if $p \in K$ and $|t| < \delta$, then:*

$$\|\exp(X(t)) \circ \exp(Y(t))(p) - \exp(H_N(X(t), Y(t)))(p)\| \leq C|t|^{N+1}.$$

We are now in a position to prove formula (1.7), that we used in the proof of Lemma 1.1. Let $X_1, \ldots, X_m$ be m elements of $V F(U)$. For every element $I \in \mathscr{L}(1, \ldots, m)$, we define the local diffeomorphisms ϕ_t^I on U by induction on the length $|I|$ of I. Let $\phi_t^i = \exp(t X_i)$ and set, if $I = [I_1, I_2]$,

$$\phi_t^I = \phi_{-t}^{I_2} \circ \phi_{-t}^{I_1} \circ \phi_t^{I_2} \circ \phi_t^{I_1}.$$

Proposition A.1 *Let $K \subset U$ be a compact and $I \in \mathscr{L}(1, \ldots, m)$. There exist two positive constants δ, C such that, if $p \in K$ and $|t| < \delta$, then*

$$\left\|\phi_t^I(p) - p - t^{|I|} X_I(p)\right\| \leq C|t|^{|I|+1}. \tag{A.3}$$

Proof For $\delta > 0$ small enough, the mapping $(p, t) \mapsto \phi_t^I(p)$ is defined and C^∞ on $K \times (-\delta, \delta)$. As a consequence, we are reduced to prove (A.3) for a fixed $p \in K$.

When $|I| = 1$, $\phi_t^I(p) = \exp(t X_i)(p)$ for some $i \in \{1, \ldots, m\}$, which is equal to $p + t X_i(p) + O(|t|^2)$.

Now, let I be an element of $\mathscr{L}(1, \ldots, m)$ and $N > |I|$ an integer. Corollary A.1 implies that $\phi_t^I(p) = \exp(H_N^I(t))(p) + O(t^{N+1})$ where the series $H_N^I(t)$ is defined by induction on the length $|I|$: if $I = i$, then $H_N^I(t) = t X_i$, and if $I = [I_1, I_2]$, then

$$H_N^I(t) = H_N(H_N(H_N^{I_1}(t), H_N^{I_2}(t)), H_N(-H_N^{I_1}(t), -H_N^{I_2}(t))).$$

Applying (A.1) iteratively, we can write $H_N^I(t) = t^{|I|} X_I + t^{|I|+1} R_I(t)$, the latter term being a one-parameter vector field. As a consequence,

$$\phi_t^I(p) = p + t^{|I|} X_I(p) + \text{terms of degree greater than } |I| + 1,$$

which completes the proof. $\qquad\square$

Remark A.1 Assume that $X_1, \ldots, X_m$ are real analytic vector fields which generate a nilpotent Lie algebra of step r. In this case, for any pair of vector fields X, Y in $Lie(X_1, \ldots, X_m)$ and any $t \in \mathbb{R}$, we have $H_N(X, Y) = H_r(X, Y)$ whenever $N \geq r$. Thus, following Lemma A.1, $\exp(t X) \circ \exp(t Y)(p)$ and $\exp(H_r(t Y, t X))(p)$ have the same Taylor expansion at $t = 0$. Since both are analytic functions of t they are equal for t small enough,

$$\exp(t X) \circ \exp(t Y)(p) = \exp(H_N(t Y, t X))(p).$$

Using this property in the proof of Proposition A.1, one proves that (A.3) can be replaced by the following equality,

$$\phi_t^I(p) = \exp\left(t^{|I|}X_I + t^{|I|+1}R_I(t)\right) = \exp\left(t^{|I|+1}R_I'(t)\right) \circ \exp\left(t^{|I|}X_I\right),$$

where $R_I(t)$ and $R_I'(t)$ are linear combinations of brackets X_J with $|I| < |J| \le r$.

A.2 Push-Forward Formula

Given two vector fields $X, Y \in VF(U)$, we write $(\mathrm{ad}X)Y$ for $[X, Y]$, $(\mathrm{ad}X)^2Y$ for $(\mathrm{ad}X)((\mathrm{ad}X)Y)$, etc.

Proposition A.2 *Let $K \subset U$ be a compact, N a positive integer, and X, Y, $Y_1, \ldots, Y_\ell$ vector fields in $VF(U)$. There exist two positive constants δ, C such that, if $p \in K$, $t \in \mathbb{R}$ and $z \in \mathbb{R}^\ell$ satisfy $|t| < \delta$ and $\|z\| < \delta$, then*

$$\left\| \exp(tY)_* X(p) - \sum_{k=0}^{N} \frac{t^k}{k!}(\mathrm{ad}Y)^k X(p) \right\| \le C|t|^{N+1},$$

$$\left\| \exp(\sum_{i=1}^{\ell} z_i Y_i)_* X(p) - \sum_{k=0}^{N} \frac{1}{k!}\left(\mathrm{ad}\sum_{i=1}^{\ell} z_i Y_i \right)^k X(p) \right\| \le C\|z\|^{N+1},$$

where $\exp(tY)_* X = d(\exp(tY)) \circ X \circ \exp(-tY)$ denotes the push-forward of the vector field X by the diffeomorphism $\exp(tY)$.

Proof Let us begin with the first inequality. Set $\phi_p(t) = \exp(tY)_* X(p)$. For $\delta > 0$ small enough, the mapping $(p, t) \mapsto \phi_p(t)$ is defined and C^∞ on $K \times (-\delta, \delta)$. As a consequence, there exists a constant $C > 0$ such that, for every $p \in K$ and $|t| < \delta$, we have

$$\left\| \phi_p(t) - \sum_{k=0}^{N} \frac{t^k}{k!} \frac{d^k \phi_p}{dt^k}(0) \right\| \le C|t|^{N+1}.$$

It remains to prove that $\frac{d^k \phi_p}{dt^k}(0) = (\mathrm{ad}Y)^k X(p)$ for any integer k. Note first that $\phi_p(0) = X(p)$, and that

$$\frac{d\phi_p}{dt}(0) = \frac{d}{dt}\left[\exp(tY)_* X\right]\big|_{t=0}(p)$$

is by definition equal to $-L_Y X(p)$, where $L_Y X$ is the Lie derivative of X along Y (see for instance [1]). Since $L_Y X(p) = (\mathrm{ad}Y)X(p)$, the cases $k = 0$ and $k = 1$ are done.

We need now to compute $\frac{d\phi_p}{dt}(t)$ at $t \neq 0$, or equivalently $\frac{d\phi_p(t+s)}{ds}$ at $s = 0$. Writing $\phi_p(t+s)$ as $\exp(tY)_* \exp(sY)_* X(p)$, we have

$$\frac{d\phi_p}{dt}(t) = \frac{d\phi_p(t+s)}{ds}\Big|_{s=0} = \exp(tY)_* \frac{d}{ds}\big[\exp(sY)_* X\big]\Big|_{s=0}(p)$$
$$= \exp(tY)_*((\mathrm{ad}\,Y)X)(p).$$

This derivative has the same form as $\phi_p(t)$, X being replaced by $(\mathrm{ad}\,Y)X$. Iterating the argument above, we obtain by induction $\frac{d^k\phi_p}{dt^k}(0) = (\mathrm{ad}\,Y)^k X(p)$, and the first inequality of the proposition is proved.

As for the second inequality, the same reasoning applies and we only need to compute the partial derivatives at $0 \in \mathbb{R}^\ell$ of the function $\tilde{\phi}(z) = \exp(\sum_i z_i Y_i)_* X(p)$. This can be done as in the proof of Lemma A.2. The proposition follows. $\square$

References

1. Boothby, W.: An Introduction to Differentiable Manifolds and Riemannian Geometry. Academic Press, Massachusetts (1986)
2. Bourbaki, N.: Groupes et Algèbres de Lie. Hermann, Paris, (1972)

Appendix B
The Different Systems of Privileged Coordinates

This appendix is devoted to the proof that the examples of coordinates introduced in Sect. 2.1.2 are actually privileged coordinates.

B.1 Canonical Coordinates of the Second Kind

Let $p \in M$ and let us choose an adapted frame at p, i.e. a family of vector fields $Y_1, \ldots, Y_n$ such that

$$\begin{cases} Y_1(p), \ldots, Y_n(p) \text{ is a basis of } T_p M, \\ Y_i \in \Delta^{w_i}, \ i = 1, \ldots, n. \end{cases}$$

The map

$$\phi \colon (z_1, \ldots, z_n) \mapsto \exp(z_n Y_n) \circ \cdots \circ \exp(z_1 Y_1)(p)$$

is a local diffeomorphism near $0 \in \mathbb{R}^n$ and its inverse defines some coordinates called *canonical coordinates of the second kind near p*.

The following result is due to Hermes [1].

Lemma B.1 *Canonical coordinates of the second kind are privileged at p.*

For sake of simplicity, we will write the compositions of maps as products; for instance, we write

$$\phi(z) = \exp(z_1 Y_1) \ldots \exp(z_n Y_n)(p).$$

Proof First, let us recall that ϕ is a local diffeomorphism at $z = 0$ because its differential at 0 is an isomorphism. This results from

© The Author(s) 2014

F. Jean, *Control of Nonholonomic Systems: From Sub-Riemannian Geometry to Motion Planning*, SpringerBriefs in Mathematics, DOI 10.1007/978-3-319-08690-3

$$\frac{\partial \phi}{\partial z_i}(0) = \frac{d}{dt}\left(\phi(0,\ldots,t,\ldots,0)\right)\Big|_{t=0} = \frac{d}{dt}\left(\exp(tY_i)(p)\right)\Big|_{t=0} = Y_i(p),$$

for $i = 1,\ldots,n$. This computation also reads as $\phi_* \frac{\partial}{\partial z_i}(p) = Y_i(p)$, which implies $Y_i z_i(p) = 1$ (as in Sect. 2.1.1, $Y_i z_i$ denotes the Lie derivative of the function z_i along the vector field Y_i). Hence the order of z_i at p is not greater than w_i.

It remains to show that the order of z_i at p is at least w_i for each $i = 1,\ldots,n$. This is a direct consequence of the following assertion.

Claim Let X be one of the vector fields $X_1,\ldots,X_m$. Then, for $i = 1,\ldots,n$, the Taylor expansion at $z = 0$ of the function $a_i(z) = Xz_i(\phi(z))$ is a sum of homogeneous polynomials in the coordinates z of weighted degree $\geq w_i - 1$.

From the very definition of $a_i(z)$, we have

$$X(\phi(z)) = \sum_{i=1}^{n} a_i(z) \frac{\partial \phi}{\partial z_i}(z). \tag{B.1}$$

Given z, let φ be the diffeomorphism defined on a neighbourhood of p by $\varphi(q) = \exp(z_1 Y_1)\ldots\exp(z_n Y_n)(q)$. In particular, $\varphi(p) = \phi(z)$. In order to obtain an equality in $T_p M$, we apply the isomorphism $(d\varphi_p)^{-1}$ to both sides of (B.1), and we get

$$(\varphi^{-1})_* X(p) = \sum_{i=1}^{n} a_i(z) (\varphi^{-1})_* \frac{\partial}{\partial z_i}(p).$$

This equality is of the form $W = \sum_{i=1}^{n} a_i V_i$, where the vectors $W = W(z)$ and $V_i = V_i(z)$, $i = 1,\ldots,n$, belong to $T_p M$. If we denote by $b = b(z) \in \mathbb{R}^n$ the coordinates of W in the basis $(Y_1(p),\ldots,Y_n(p))$ of $T_p M$, and by $P = P(z)$ the $(n \times n)$-matrix of the coordinates of $V_1,\ldots,V_n$ in the same basis, then the vector $a(z) = (a_1(z),\ldots,a_n(z))$ appears as the solution of $Pa = b$.

Note first that $P(0)$ equals the identity matrix Id. Therefore both matrices $P(z)$ and $P(z)^{-1}$ are equal to Id+homogeneous terms of positive degree. Hence the Taylor expansion of $a_i(z)$ and the one of $b_i(z)$ have the same homogeneous terms of lower degree.

On the other hand, since $\varphi^{-1} = \exp(-z_n Y_n)\ldots\exp(-z_1 Y_1)$, we have

$$W(z) = \exp(-z_n Y_n)_* \ldots \exp(-z_1 Y_1)_* X(p).$$

Let us choose an integer N bigger than all the weights w_i, and apply Proposition A.2 to $\exp(-z_1 Y_1)_* X$, then to $\exp(-z_2 Y_2)_*(\mathrm{ad} Y_1)^{l_1} X$, and so on,

$$W(z) = \exp(-z_n Y_n)_* \ldots \exp(-z_2 Y_2)_* \sum_{l_1=0}^{N} \frac{(-z_1)^{l_1}}{l_1!} (\mathrm{ad}\, Y_1)^{l_1} X(p) + O(|z_1|^{N+1})$$

$$\vdots$$

$$= \sum_{l_1,\ldots,l_n=0}^{N} \frac{(-z_1)^{l_1}}{l_1!} \cdots \frac{(-z_n)^{l_n}}{l_n!} (\mathrm{ad}\, Y_n)^{l_n} \ldots (\mathrm{ad}\, Y_1)^{l_1} X(p) + O(|z|^{N+1}).$$

Hence every coordinate $b_i(z)$ of $W(z)$ satisfies

$$b_i(z) = \sum_{l_1,\ldots,l_n=0}^{N} \frac{(-z_1)^{l_1}}{l_1!} \cdots \frac{(-z_n)^{l_n}}{l_n!} \beta_i^l + O(|z|^{N+1}), \tag{B.2}$$

where β_i^l denotes the ith coordinate of the vector $(\mathrm{ad}\, Y_n)^{l_n} \ldots (\mathrm{ad}\, Y_1)^{l_1} X(p)$ in the basis $(Y_1(p), \ldots, Y_n(p))$. The latter vector belongs to $\Delta^w(p)$, where $w = 1 + l_1 w_1 + \cdots + l_n w_n$ (recall that $X \in \Delta^1$ and $Y_i \in \Delta^{w_i}$). Since $(Y_1, \ldots, Y_n)$ is an adapted frame at p, β_i^l is zero when $1 + l_1 w_1 + \cdots + l_n w_n < w_i$. It follows that $b_i(z)$ – and then $a_i(z)$ – contains only homogeneous terms of weighted degree greater than or equal to $w_i - 1$. This ends the proofs of both the claim and the lemma. $\qquad\square$

B.2 Canonical Coordinates of the First Kind

Let $p \in M$ and $Y_1, \ldots, Y_n$ an adapted frame at p. The map

$$\widetilde{\phi} \colon (z_1, \ldots, z_n) \mapsto \exp(z_1 Y_1 + \cdots + z_n Y_n)(p)$$

is a local diffeomorphism near $0 \in \mathbb{R}^n$ since its differential at 0 is an isomorphism. This results from

$$\frac{\partial \widetilde{\phi}}{\partial z_i}(0) = \frac{d}{dt}\left(\widetilde{\phi}(0, \ldots, t, \ldots, 0)\right)\Big|_{t=0} = \frac{d}{dt}\left(\exp(t Y_i)(p)\right)\Big|_{t=0} = Y_i(p),$$

for $i = 1, \ldots, n$. The inverse of $\widetilde{\phi}$ defines some local coordinates near p called *canonical coordinates of the first kind*.

Lemma B.2 *Canonical coordinates of the first kind are privileged at p.*

The first proof of this lemma appeared in [2], with a different formulation. The proof we present here is rather different.

Proof The proof follows exactly the same lines as the one of Lemma B.1, replacing ϕ by $\widetilde{\phi}$, and φ by $\widetilde{\varphi} = \exp(\sum_j z_j Y_j)$. We are left to compute the coordinates $\widetilde{b}_i(z)$, $i = 1, \ldots, n$, of the vector $\widetilde{W}(z) = (\widetilde{\varphi}^{-1})_* X(p)$ in the basis $(Y_1(p), \ldots, Y_n(p))$ of $T_p M$. It results directly from Proposition A.2 that

$$
\widetilde{W}(z) = \sum_{k=0}^{N} \frac{1}{k!} \left(\mathrm{ad} \sum_{i=1}^{\ell} z_j Y_j \right)^k X(p) + O(|z|^{N+1}),
$$

$$
= \sum_{k=0}^{N} \sum_{l_1 + \cdots + l_n = k} a_l z_1^{l_1} \ldots z_n^{l_n} Z_l(p) + O(|z|^{N+1}),
$$

where Z_l belongs to $\Delta^w(p)$, with $w = 1 + l_1 w_1 + \cdots + l_n w_n$. Thus every coordinate $\widetilde{b}_i(z)$ of $\widetilde{W}(z)$ satisfies

$$
\widetilde{b}_i(z) = \sum_{k=0}^{N} \sum_{l_1 + \cdots + l_n = k} a_l z_1^{l_1} \ldots z_n^{l_n} \widetilde{\beta}_i^l + O(|z|^{N+1}),
$$

$\widetilde{\beta}_i^l$ being the ith coordinate of Z_l in the basis $(Y_1(p), \ldots, Y_n(p))$. This expression is similar to (B.2), and the same conclusion follows. $\qquad\square$

B.3 Algebraic Coordinates

Let us recall the construction of the algebraic coordinates $(z_1, \ldots, z_n)$ given in page 23. Let $Y_1, \ldots, Y_n$ be an adapted frame at p, and $(y_1, \ldots, y_n)$ be local coordinates centered at p such that $\partial_{y_i}|_p = Y_i(p)$. For $j = 1, \ldots, n$, we set

$$
z_j = y_j - \sum_{k=2}^{w_j - 1} h_k(y_1, \ldots, y_{j-1}), \tag{B.3}
$$

where, for $k = 2, \ldots, w_j - 1$,

$$
h_k(y_1, \ldots, y_{j-1}) = \sum_{\substack{|\alpha| = k \\ w(\alpha) < w_j}} Y_1^{\alpha_1} \ldots Y_{j-1}^{\alpha_{j-1}} \left(y_j - \sum_{q=2}^{k-1} h_q(y) \right)(p) \frac{y_1^{\alpha_1}}{\alpha_1!} \ldots \frac{y_{j-1}^{\alpha_{j-1}}}{\alpha_{j-1}!},
$$

with $|\alpha| = \alpha_1 + \cdots + \alpha_n$.

Lemma B.3 *The algebraic coordinates* $(z_1, \ldots, z_n)$ *are privileged at* p.

The proof of the lemma is based on the following result.

Lemma B.4 *A function f is of order $\geq s$ at p if and only if*

$$(Y_1^{\alpha_1} \cdots Y_n^{\alpha_n} f)(p) = 0$$

for all α such that $w(\alpha) < s$.

Proof Let f be a function of order $\geq s$ at p. Using the rules (2.1), we have $\mathrm{ord}_p(Y_i) \geq -w_i$ for $i = 1, \ldots, n$, and hence $\mathrm{ord}_p(Y_1^{\alpha_1} \cdots Y_n^{\alpha_n}) > -s$ for every $\alpha = (\alpha_1, \ldots, \alpha_n)$ such that $w(\alpha) < s$. Consequently, for such an α the iterated derivative $Y_1^{\alpha_1} \cdots Y_n^{\alpha_n} f$ is a function of positive order, and so vanishes at p.

Conversely, let f be a function of order $< s$ at p. We introduce the canonical coordinates of the second kind $(x_1, \ldots, x_n)$ defined by means of the adapted basis $Y_1, \ldots, Y_n$. Proposition 2.2 implies that there exists α such that $w(\alpha) = \mathrm{ord}_p(f) < s$ and $(\partial_{x_1}^{\alpha_1} \cdots \partial_{x_n}^{\alpha_n} f)(p) \neq 0$. Moreover, every vector field Y_i, $i = 1, \ldots, n$, writes in coordinates x as

$$\sum_{j=1}^{n} Y_i^j(x)\partial_{x_j}, \quad \text{where } \mathrm{ord}_p(Y_i^j) \geq w_j - w_i.$$

There also holds $Y_i^j(0) = \delta_{ij}$ since $Y_i(p) = \partial_{x_i}$. As a consequence,

$$Y_1^{\alpha_1} \cdots Y_n^{\alpha_n}(p) = \partial_{x_1}^{\alpha_1} \cdots \partial_{x_n}^{\alpha_n}(p) + \sum_{w(\beta)<w(\alpha)} a_\beta \partial_{x_1}^{\beta_1} \cdots \partial_{x_n}^{\beta_n}(p),$$

and thus $(Y_1^{\alpha_1} \cdots Y_n^{\alpha_n} f)(p) = (\partial_{x_1}^{\alpha_1} \cdots \partial_{x_n}^{\alpha_n} f)(p) \neq 0$ since $w(\alpha) = \mathrm{ord}_p(f)$. This ends the proof. $\qquad\square$

Proof (of Lemma B.3) Let $i \in \{1, \ldots, n\}$. Note first that $Y_i z_i(p) = Y_i y_i(p) = 1$, which implies $\mathrm{ord}_p(z_i) \leq w_i$. It remains to show that $\mathrm{ord}_p(z_i) \geq w_i$. For this we will use the criterion of Lemma B.4.

Let α such that $w(\alpha) < w_i$ (and so $|\alpha| < w_i$). Using the expression (B.3) of z_i, we obtain

$$Y_1^{\alpha_1} \ldots Y_n^{\alpha_n} z_i = Y_1^{\alpha_1} \ldots Y_n^{\alpha_n} \left(y_i - \sum_{k=2}^{w_i-1} h_k(y) \right)$$

$$= Y_1^{\alpha_1} \ldots Y_n^{\alpha_n} \left(y_i - \sum_{k=2}^{|\alpha|-1} h_k(y) \right) - Y_1^{\alpha_1} \ldots Y_n^{\alpha_n} \left(\sum_{k=|\alpha|}^{w_i-1} h_k(y) \right).$$

$$\tag{B.4}$$

The functions h_k are given by

$$h_k(y) = \sum_{\substack{|\beta|=k \\ w(\beta)<w_i}} Y_1^{\beta_1} \ldots Y_{i-1}^{\beta_{i-1}} \left(y_i - \sum_{q=2}^{k-1} h_q(y) \right)(p) \, \frac{y_1^{\beta_1}}{\beta_1!} \ldots \frac{y_{i-1}^{\beta_{i-1}}}{\beta_{i-1}!}.$$

Therefore, we clearly have $\left(Y_1^{\alpha_1} \ldots Y_n^{\alpha_n} h_k\right)(p) = 0$ if $k > |\alpha|$, and

$$\left(Y_1^{\alpha_1} \ldots Y_n^{\alpha_n} h_{|\alpha|}\right)(p) = Y_1^{\alpha_1} \ldots Y_n^{\alpha_n} \left(y_i - \sum_{k=2}^{|\alpha|-1} h_k(y) \right)(p).$$

Plugging this expression into (B.4), we obtain $(Y_1^{\alpha_1} \ldots Y_n^{\alpha_n} z_i)(p) = 0$, which ends the proof. $\qquad\square$

References

1. Hermes, H.: Nilpotent and high-order approximations of vector field systems. SIAM Rev., **33**(2), 238–264 (1991)
2. Rothschild, L.P., Stein E.M.: Hypoelliptic differential operators and nilpotent groups. Acta. Math., **137**, 247–320 (1976)

MIX
Papier aus verantwortungsvollen Quellen
Paper from responsible sources
FSC® C105338

If you have any concerns about our products,
you can contact us on
ProductSafety@springernature.com

In case Publisher is established outside the EU,
the EU authorized representative is:
Springer Nature Customer Service Center GmbH
Europaplatz 3, 69115 Heidelberg, Germany

Printed by Libri Plureos GmbH
in Hamburg, Germany